Writing for the Web

MW00417003

Writing for the Web unites theory, technology, and practice to explore writing and hypertext for Web site creation. It integrates such key topics as XHTML/CSS coding, writing (prose) for the Web, the rhetorical needs of the audience, theories of hypertext, usability and architecture, and the basics of Web site design and technology. Presenting information in digestible parts, this text enables students to write and construct realistic and manageable Web sites with a strong theoretical understanding of how online texts communicate to audiences.

Key features of the book include:

- Screenshots of contemporary Web sites that will allow students to understand how writing for and linking to other layers of a Web site should work.
- Flow charts that describe how Web site architecture and navigation works.
- Parsing exercises in which students break down information into subsets to demonstrate how Web site architecture can be usable and scalable.
- Detailed step-by-step descriptions of how to use basic technologies such as file transfer protocols (FTP).
- Hands-on projects for students to engage in that allow them to connect the various components in the text.
- A companion website with downloadable code and additional pedagogical features: www.routledge.com/cw/applen

Writing for the Web prepares students to work in professional roles, as it facilitates understanding of architecture and arrangement of written content of an organization's texts.

J.D. Applen is an associate professor of English at the University of Central Florida. He is interested in writing and rhetoric, technical communication, the literature of science and technology, and the rhetoric of hypertext, digital archives, science, and the environment.

Writing for the Web

Composing, Coding, and Constructing Web sites

J.D. Applen

Routledge
Taylor & Francis Group

NEW YORK AND LONDON

Please visit the companion website for this book at www.routledge.com/cw/applen

First published 2013
by Routledge
711 Third Avenue, New York, NY 10017

Simultaneously published in the UK
by Routledge
2 Park Square, Milton Park, Abingdon, Oxon OX14 4RN

Routledge is an imprint of the Taylor & Francis Group, an informa business

© 2013 Taylor & Francis

The right of J.D. Applen to be identified as author of this work has been asserted by him in accordance with sections 77 and 78 of the Copyright, Designs and Patents Act 1988.

Library of Congress Cataloging in Publication Data
Applen, J. D.
Writing for the web : composing, coding, and constructing web sites / by J.D. Applen.
    pages cm
1. Web sites—Design. 2. Online authorship. I. Title.
TK5105.888.A63 2013
006.7—dc23   2012038732

ISBN: 978-0-415-88325-2 (hbk)
ISBN: 978-0-415-88326-9 (pbk)
ISBN: 978-0-203-84525-7 (ebk)

Typeset in Bembo
by Cenveo Publisher Services

Certified Sourcing
www.sfiprogram.org
SFI-00453

Printed and bound in the United States of America
by Edwards Brothers, Inc.

Contents

Figures

Tables

Preface

This book describes, demonstrates how to apply, and integrates the writing, organizing, and technical skills one needs to produce informative Web sites. To this end, we need to assume the following roles:

- **A competent media theorist**—One who recognizes in what manner a hypertext document both communicates and shapes our understanding of the information it presents to us. This requires a critical knowledge of the communication practices that have been in play in our civilization over the last twenty-five hundred years and how they compare to the skill of writing.
- **A competent technician**—One who knows how HTML and CSS work to create electronic documents and who can construct Web sites from scratch with them. All competent writers can extend the reach of what they write by knowing how to apply these technologies. It is important not to be intimidated by these technologies because they are really not that difficult to learn; good writers can become skilled coders and do not always need to depend on other "technicians" or HTML/CSS editors.
- **A competent rhetorician**—One who can identify the rhetorical effects of all texts and how these effects can inform a writer or Web site architect's decisions about her own work and allow her to become a discerning consumer of other information found on the World Wide Web. A working knowledge of rhetoric supports all of the other skills described in this book.
- **A competent writer**—One who understands what clear and effective writing is and how it can be applied to documents found on the World Wide Web. Being able to encode someone else's writing and other texts using HTML and CSS for a Web site is one thing, but being able to write well, which means being able to research and describe your own ideas or document the ideas of others with whom you work, is equally important. Additionally, breaking down this information and then organizing it into hyperlinked bodies of text is a role the writer needs to assume.

There are many insightful media theorists, able HTML/CSS technicians, insightful rhetoricians, and lucid writers, but to be able to completely understand and be an effective producer of online information we have to accept the responsibility of becoming competent in all these areas.

We often hear the word "literacy" used to describe certain competencies we need to function in today's economy. According to the Association of College and Research Libraries (ACRL), "information literacy" means knowing how to select and use appropriate databases, implement a search strategy, and evaluate different viewpoints—skills that educated citizens need to be successful. Information literacy can describe the practice of reading a passage in a book and culling the meaning from it, which is the more traditional understanding of what it means to be literate. Acquiring these skills is not always easy, but is immensely rewarding.

Sometimes information literacy is thought of as being in the same vein as information technology skills—they are not identical, but there is some overlap between them. Being "computer literate" means that we know how hardware and software work in today's communication technologies, but "fluency with technology" has to do with our using technology to find information and use it appropriately (ACRL). Certainly, we do need some technical skills with computers to begin finding information before we can begin to evaluate it.

"Digital literacy" is another term that is being used by professionals these days. It conflates some of the elements of the concepts described above, but also extends them. A good working definition of digital literacy has been proposed by Rachel Spilka:

> Theory and practice that focus on use of digital technology, including the ability to read, write, and communicate using digital technology, the ability to think critically about digital technology, and consideration of social, cultural, political, and educational values associated with those activities.
>
> Spilka (8)

To become digitally literate, we are asked to be writers and communicators who can use digital technologies and be able to "think critically" about them across a broad range of concerns. Understanding the "social, cultural, political, and educational values" requires that we examine how communication technologies present and shape information. It is not about the content digital technologies deliver, but about how the machines themselves actually affect and perhaps alter our understanding of the content. To do this, we need to understand the history of communication technology and how it has been theorized so we can become more critically engaged as we work to produce and present information.

There are many complex Web sites in operation today that are beautiful, competently constructed, and invite their audiences to interact with the material on them in ingenious ways, but these Web sites are built by large teams of professionals—computer scientists, information technologists, technical communicators, and graphic artists—and these Web sites are developed over months and years. To be able to teach any one student to be able to build such Web sites would be unrealistic, and there is not a single book that could teach this. However, the basis or *sine qua non* for these Web sites is HTML and CSS, and learning these technologies compels us to understand the essential styling, layout, and organizing strategies of all Web sites.

If you know how HTML works with CSS, you are in effect an electronic layout specialist and are carrying on the work of the typesetters of the last six centuries. In addition, just performing exercises using HTML tags that allow you to link different parts of a larger online document compels you to internalize the nature of hypertext as it contrasts to more traditional linear texts, thus enhancing the digital literacy skills described above. The rhetorical features of hypertext require that you break information down and connect it in another way because it communicates information differently than traditional print texts. This is empowering, and it would enable a student to prepare for work as a member of a large production team in the field because the student would understand the architecture and arrangement of written content in an organization's texts.

HTML and CSS code are presented in this book in a way that people across all disciplines can understand. Every coded example is explained, which enables students to reconfigure the written texts they have produced for the Web environment. Additionally, there are accompanying screenshots of this code so students can know whether or not they are entering and manipulating it correctly. Some of the coding is relatively sophisticated and care has gone into explaining it with an interdisciplinary audience in mind, especially the way CSS is used for IDs, classes, and overall layout.

HTML and CSS, while explained in their separate chapters, are integrated into two final Web site assignments, one with two versions; there is a personal Web site and two informational Web sites, one based on breadcrumbs and one on drop-down menus. All of the technology explained in Chapters 2 and 3 finds its way into these Web sites, which is a departure from many books on HTML and CSS, which break them down into their specific technical features but do not show how to integrate them in a larger Web site project. With an understanding of the technologies in Chapters 2 and 3, students and instructors can manipulate the code for the projects explained in Chapter 6 for their own needs. Most importantly, the HTML and CSS technology in this last chapter supports the transfer of the written word to online environments.

This book was written in an effort to explain how to write, organize, and then use HTML and CSS coding to best present our writing in online environments. To this end, classical and contemporary rhetoric(s) have been explicated in a way that students can understand, taking as examples contemporary situations. These rhetorical tools can be applied to Web sites and online writing. The history of communication has been carefully described so students can understand just how media has changed the manner in which we think and communicate. This has been presented with the knowledge that students are interested in these matters—matters such as the difference between the oral communication of the ancients and the electronic communication of today—and can benefit from this applied theoretical perspective.

I would like to thank my colleagues in the University of Central Florida's College of Arts and Humanities for awarding me a sabbatical that enabled me to begin this project. Linda Bathgate of Routledge deserves credit for her skill as a Publisher, most notably for her quick and thorough responses to any question I had regarding this manuscript, and I appreciate Julia Sammaritano's contributions as Senior Editorial Assistant. Additionally, I want to acknowledge the support of my colleagues in the literature and technical communication tracks in the Department of English: Pat Angley, Paul Dombrowski, Madelyn Flammia, Dan Jones, Lisa Logan, Rudy McDaniel, and Patrick Murphy. Most of all, I would like to thank my family members for their support and encouragement.

J.D. Applen
Orlando, Florida

Works cited

Association of College and Research Libraries. "Information Literacy Competency Standards for Higher Education." Web. December 2010.

Spilka, Rachel. "Introduction." In *Digital Literacy for Technical Communication: 21st Century Theory and Practice.* Ed. Rachel Spilka. New York: Routledge, 2010, 1–18. Print.

About the Author

J.D. Applen is an associate professor of English at the University of Central Florida. He is interested in writing and rhetoric, technical communication, the literature of science and technology, and the rhetoric of hypertext, digital archives, science, and the environment.

Copyright Acknowledgments

Adobe product screenshots used with the permission of Adobe Systems Incorporated.

"Brittney Adams" personal Web site screenshots and HTML/CSS coding and text used with the permission of Brittney Adams.

Florida map used with the permission of the Florida Center for Instructional Technology.

Mayo Clinic screenshots used with the permission of Mayo Clinic.

Microsoft Internet Explorer and Notepad screenshots used with the permission of Microsoft.

Mozilla Firefox screenshots used with permission and in accordance with Mozilla permissions policy.

FireFTP screenshots used with permission of Mime Čuvalo.

Tim Berners-Lee image and W3C logo used with permission of the World Wide Web Consortium.

"We Can Do It" image and the Gettysburg Address used with the permission of the National Archives and Records Administration.

National Park Service (NPS) Web site screenshots used with permission of the NPS.

Chapter 1

Old media, new media, and knowledge

Chapter overview

The purpose of this chapter is to raise awareness of how the major communication media that have been in use in the last 2500 years allow us to shape and receive information and knowledge. The spoken word or oral communication of the ancients, the written word of yesterday and today, and the electronic word that we encounter in using contemporary technologies all have distinctive capacities to convey ideas to audiences and affect the speaker's or writer's ability to understand the ideas she or he works to convey. This is important because as writers and information architects we need to be mindful of the idea that the technologies we employ are not neutral.

Speaking, writing, and literacy

The most compelling idea in Walter Ong's *Orality and Literacy* is that "More than any single convention, writing has transformed human consciousness" (77). "Human consciousness," for humans at least, is just about everything, and when the advent of writing is identified as the most significant catalyst for the development of human consciousness we should stop and reflect on this idea. All the things that we think about and are aware of might be a good way of understanding what constitutes human consciousness.

In today's culture we are told by people who are trying to sell us new technologies that the "information age" is the most important period in the history of humankind, but communication technology was not invented in the 1980s. We need to recognize that writing too is a technology, even though it might be hard to imagine this today, given that pencils and paper seem like primitive artifacts when compared to computers and network technologies. Writing is a technology because it exists outside of our minds; humans needed to invent an alphabet and something to write with and on so they could record and see what they were thinking, and this was an extraordinary achievement. However, what contemporary communication technologies have given us is also remarkable. Anyone in the world can send "mail" to anyone else in the world instantaneously, and we can present ourselves and the organizations we work for across the globe via our Web sites with great ease and efficiency.

"Orality" is the word Walter Ong uses to characterize speech, and "literacy" is the word he uses for written language. Ong not only characterizes the nature of orality and literacy, more importantly he explains how both of these methods of communication affect the way we think and understand our worlds.

Before we began writing things down, approximately 2500 years ago, communication was all oral and thus based in sound—which is obvious, but according to Ong, the sound of language is still one of its prominent characteristics today, even though much of what we communicate with is written discourse. Ong takes it further when he states that "thought itself relates in an altogether special way to sound" (Ong 7). Perhaps we can understand this as, when we read something today, whether it is a book, a Web site, or a text message, we can "hear" the language we are reading as if it is being spoken.

The vast majority of languages being used by people today exist only in oral form. We might be surprised by this, but there are small populations of indigenous peoples in the world who have retained their oral languages without written components. The grammatical rules of "natural human languages" that are spoken and heard come into being long before these languages are studied by grammarians and linguists and then written down and taught to people in schools.

By contrast, the "grammar" or syntax of computer languages can be designed before they are used as they are based on the abstract logic of technology, rather than on the day-to-day needs of people who are working to communicate with one another (Ong 7). Usually computer languages are for relatively simple pieces of data, not abstract human concepts; describing the satisfaction we experience in the company of our friends or family or why we like the music we do cannot be reduced to a number or one-word designation found in a database. The rules or grammar of computer languages can be designed to accomplish a known set of tasks, and, once designed, can then be put into machines to carry out these tasks immediately.

Think about some of the common expressions you use that, while they have meaning, are difficult to convey logically. Go to urbandictionary.com and find an expression that can only be explained by relating it to the context it is used in, not by the actual logic of the sentence or the literal meaning of each word in the sentence or phrase. How difficult would it be to program a computer to understand these expressions and use them correctly in natural human speech? For example, how would you explain the phrase "That's how I roll"? How might it be different than "That's how I do things"?

"Natural human languages" develop over time; new words or ways to express ourselves come into play oftentimes as slang, and because of the fact that people use new words and expressions and they are understood by others in their community when they do. These new ways of speaking gain respectability as they successfully convey meaning and thus become part of the "official" language and are written down in dictionaries. For example, "google" in the sense of "to google something" started out as slang, gained widespread use, and can now be found in respectable dictionaries. It made sense to enough people in our culture that this was a good word to use, perhaps because before it was a verb it was used as a proper noun: it was the name of the most widely used search engine, Google, and "to google" allowed people to distinguish easily between just searching on any search engine and searching on Google. We will be discussing this "social construction" of language in more detail in Chapter 4.

Ong tells us that people in oral cultures do not learn like people do in written cultures. Learning in oral cultures is based on the practice of apprenticeship, where people imitate or emulate what others do as they are doing it. For example, people learn how to hunt by going on hunts and emulating what experienced hunters do. This has to happen in real time; it is not something that is read about in a book before people go out and actually do it. While oral cultures are analytic in that people in them need to work to find words and phrases to describe things and ideas, people in literate cultures, according to Ong, "study" things because written language allows them to carefully examine ideas that are arrested in words; students and scholars can keep poring over a sentence or passage that explains a thing or idea at their own pace, reflect on the meaning of each word or phrase and the sequence of words, and then come to the overall meaning of what has been written. For example, if the last sentence was not quite clear to you, you could go back and reread it and think about such things as the meaning of each word as you understand it and its relationship to the other words in the sentence, the sentence's syntax or arrangement, and how the punctuation marks served to convey meaning. You could "study" it, as Ong would say, or move on to the next sentence if you so choose.

In speech, communication is not arrested in time, and unless you can remember exactly everything that you have heard, it is more difficult to reflect on than a text that has been written down and that you can keep going back to, a text that has not changed. Ong describes the inability of members from oral cultures to understand such things as "geometrical figures, abstract categorization, formally logical reasoning processes, definitions, or even comprehensive descriptions, or articulated self-analysis, all of which derive not simply from thought itself but from text-formed thought" (55). Imagine studying concepts in chemistry such as redox equations or ideas in government such as the Bill of Rights without being able to "see it" in a language that is written down. Regarding the "study" aspect of writing, think about some of the things you have learned by imitating others and other things you have learned by reading about them. What was the difference in the learning process between the two? What kinds of things are best learned by reading versus emulating the actions of others?

Ong tells us that in oral cultures "knowledge, once acquired, had to be constantly repeated or it would be lost: fixed, formulaic thought patterns were essential for wisdom and effective administration" (23). The Greeks around the time of Plato were the first to fix oral speech in written form, and by writing things down they could move on to other things; they did not have to spend so much of their time and energy remembering what they had just heard, as people do in oral cultures: it "freed the mind for more original, more abstract thought" (Ong 24).

In an oral culture people took in literature and philosophy and other important ideas by hearing it, not reading it. In the "Phaedrus" (ironically a written text) Plato lamented the loss of the practice of oral communication because he thought that we would lose our ability to memorize things if we began writing ideas down: "For this invention [writing] will produce forget-fulness in the minds of those who learn to use it, because they will not practice their memory" (140). If something was written down, we could read it, put it down, and come back to it later if we needed to, but we would lose something by doing this. Actually, when he writes this, Plato uses the voice of Socrates, who is in dialogue with a man named Phaedrus. Plato believed it was through the give and take of such questioning and answering, or what he called "dialectic," that we can best present our ideas and demon-strate how ideas can be generated and examined. This is also called the "Socratic method," which Plato learned from Socrates. Plato believed there was something special about the dialectic method as it was dynamic, whereas writing lacked this character, and we can see this expressed in Socrates' answer to Phaedrus:

> Writing, Phaedrus, has this strange quality, and is very like painting; for the creatures of a painting stand like living beings, but if one asks them a question, they preserve a solemn silence. And so it is with written words; you might think they spoke as if they had intelligence, but if you question them, wishing to know about their sayings, they always say only one and the same thing. And every word, when once it was writ-ten, is bandied about, alike among those who understand and those who have no interest in it, and it knows not to whom to speak or not to speak; when ill-treated or unjustly reviled it always needs a father to help it; for it has no power to help or protect itself.
>
> (140–141)

Written words, however elegant and specific, just sit there on the page, and if not understood or challenged have "no power to protect themselves," while the language of two or more people exchanging ideas allows them to change and clarify their words as they make adjustments for their audiences.

Think about some of the things you have written that others have read. When they have reacted to what you wrote, you might have clarified something that you thought they misunderstood by rephrasing your words or adding to it in some way. In this way, you defended your writing, but you could not have done this if someone read what you wrote and you were not there to engage them. However, because the act of writing requires the use of a technology that exists outside of our mind, it is unchanging, fixed, and removed "from the living human lifeworld" (Ong 89). The written word lives on forever, which is ironic given this characterization, whereas words that were spoken in lively but evanescent debate, and never written down, are dead (Ong 80).

Some would say that the blogs that are today prevalent online, which are based on the written word, in some ways also model the spoken word of Plato's age. You can write something out, and others might respond to it in a way that shows that they did not really "get it" the first time or they may present challenges to you. You can come back relatively quickly with a clarification or even a refinement or modification in your thinking in a process that is akin to the dialectic method.

According to Ong, the written word also allows us to produce more "abstractly sequential, classificatory, explanatory examination of phenomena or of stated truths" (24), as a writer can take the time to go back and carefully craft her sentences until they reflect what she is really trying to say. People who come from literate culture transfer the communication patterns they have learned or "interiorized" from the practice of writing to the way they speak (Ong 57). People who have never learned to read and write would not have internalized the sentence patterns that writing allows. For example, a "writer" can more readily produce a long sentence that includes such elements as dependent clauses, independent clauses, phrases, and lists separated by commas, colons, semi-colons, and dashes, because the writer has a chance to write a sentence out, reread it and think about how it sounds and communicates, then go back and rewrite it, much as this writer has done—and has had to do—with this sentence. Someone from an oral culture cannot do this as readily as someone writing sentences out with a pencil or pen, or, even more enabling, using a word processing program which invites writing and rewriting. Because of this, a literate person's speech adopts some of the patterns that are found in writing. This does not mean that we always speak like we write, but writing does have some effect.

This is not to say that oral cultures do not produce "complex and intelligent" texts (Ong 56). Before oral cultures were studied to any extent and consequently given some respect, many scholars assumed that, as it was so "skillful," Homer's *Odyssey* must have been a written text, not spoken, reflecting their failure to see how "sophisticated" and "reflective" an oral text can be (Ong 56). Before the *Odyssey* was written down, it existed as a set of oral narratives that poets would memorize and deliver to an audience.

The methods used to memorize such large bodies of text were based on knowing the meter or required poetic rhythm in every line. Meter served as a mnemonic device, a method allowing us to remember longer texts; the repetitive rhythm of each line in an epic poem suggested to a poet the next word or phrase. We are using a mnemonic device when we are working to remember which letter in our alphabet comes before the other when we sing the "ABC song"—whether out loud or silently to ourselves—which is melodically and rhythmically identical to the song "Twinkle, Twinkle, Little Star."

How much play or improvisation existed during the time of this oral culture, before Homer "wrote down" *The Odyssey*, is not known, but Ong describes studies of living oral poets from the Balkans who recited epic poems that, while they had the same basic story and content, changed in every oral presentation (59). This allowed the poets to be "original," as they might alter some of the specific words and phrasings for each audience and situation, but still to present the "traditional materials effectively" (59). While this is oral literature, presenting it in this way is a sophisticated practice. In fact, when such poems are written down the inventive thought processes of an oral poet are lost to some extent. When an "illiterate" poet eventually learns how to write something down, "it introduces into his mind the concept of a text as controlling the narrative and thereby interferes with the oral composing processes" and thus "disables the oral poet" (Lord, quoted in Ong 59). The poetry is written down, arrested, and that is that.

In today's culture, the implicit power of the written word is evident when we go see a movie that is based on a book we have read. We will often make judgments about how the movie fails to live up to the specifics of the book as opposed to taking into consideration that the director of a movie might have to use a different medium to tell the story and thus use different methods to convey the ideas of the "original" text, the book. However, sometimes the book is just better, and sometimes mediocre books can be made into good movies.

Ong distinguishes between several kinds of orality. "Primary orality" describes the way people spoke, heard, and thought before the age of literacy. "Secondary orality" describes the speaking and listening skills of people influenced by their literate background, or how anyone who could read this sentence would speak and hear things (11). For example, if someone has been exposed to written texts with long, complex sentences, punctuation marks, and embedded phrases, he or she will oftentimes speak longer sentences that have the same kind of structure and pauses signaled by such punctuation and structure; to some extent, our speech patterns sound like written sentences. People who have never learned how to read or write the language they speak would speak differently. This kind of speaking would probably also require that those who have not been exposed to the written word have not been around people from secondary orality-like cultures and thus been influenced by hearing them. There are not many people living today who would be in this group.

Secondary orality is perhaps best exemplified by what we hear on radio and television, where people are speaking but often reading from prepared written material. Network newscasters read from teleprompters and look into the camera as they speak, seemingly establishing eye contact with us and perhaps providing the illusion of a spontaneous presentation of ideas based on the memory and inventiveness of the newscaster, but this certainly is not the case and the communication is thus different from primary orality. These presentations are perfect or near perfect in terms of expression and punctuation, but this is because they are based on carefully prepared and edited written texts. As we are influenced by texts that we hear, we work them into our own speech patterns. Secondary orality is "more deliberate and self-conscious" because it is "based permanently on the use of writing and print" (Ong 134). Technologies such as radio, television, and telephones are media for secondary orality-style discourse, and their output could not be manufactured without the existence of the written word (Ong 134).

New media and literacy

Richard Lanham believes that our responses to writing and graphics on computer screens can be likened to the way we have responded to oral discourse, as opposed to print in books, in the past (34). When we take in information from a screen, "the text becomes unfixed and interactive" and the reader can "change it," and assume the role of writer (31). Lanham tells us that the "electronic word" that we come into contact with on our computer screens affects our perception or "human sensorium" differently than the traditional printed word. Traditional print texts have "No pictures; no color; strict order of left to right then down one line; no type changes; no interaction; no revision" (34), and thus make the reader feel that the writer's textual meaning is not open to any interaction.

To reach this conclusion Lanham starts with an idea of Eric Havelock's, that literate cultures have to be based on alphabets that are easy to learn and which then become transparent; the letters in the alphabet and the words on the page that are used to construct sentences are not seen but seen through. Readers never question what letters are. Instead they just decode them as they read to get the meaning of the words of the writer; they do not stop to "notice the size and shape of the letters" (33).

When we view text in online environments, we are more aware of the letters in alphabets; they are "opaque" and we see them for what they are and do not just look through them. This is in part due to the fact that we know how to manipulate letters and words in electronic environments, and thus can see them change before us; they are not just laid out in an unchanging format, as in a book. For example, we know how to choose font sizes as we have probably done some desktop publishing or just done a lot of writing and reading while looking at a computer screen where we can make adjustments to the texts we input or read. This was not the case when the most we might have done is write a paper or letter on a typewriter, or just been a reader of books and written out our ideas with a pen. We sense that text on a screen is more dynamic and changeable, that we can question it and interact with it, and this is akin to the give and take between participants in oral discourse, the dialectic method. Think of the example of blogs in the previous section: they are also dynamic and changeable.

One of the criteria that Lev Manovich uses to describe new media technologies is modularity. Units of information are modular if they do not change their "separate identities" and retain their "independence" when they are used to construct larger bodies of information (Manovich 30). For example, a media clip that has been placed in a Word document that has been edited with another software program can be thought of as an "object" that retains its independent character. This is also true of graphics that are placed on Web sites using HTML. The binary information that constitutes a jpeg or gif graphic stays unchanged and thus independent of the HTML technology used to place it on a Web site; they are separate new media objects that can be used, but not modified, to construct the whole of a Web site. HTML can be used to put a graphic on a Web site and position it relative to other information in each file, but not to alter it. It is a modular unit.

The separate objects that make up the World Wide Web are also modular. They are all connected to one another, yet they retain their separate characters. For example, the manager of one Web site can decide to link her Web site to another Web site, but she cannot make any changes to the Web site that she is linking to and thus it exists as an independent entity. Manovich describes a technology that can copy a graphic from one Web site for use on another without identifying the initial Web site where the graphic is referenced, thus demonstrating the "fundamentally discrete and non-hierarchical nature of the Web" (31). Once the graphic is placed on a Web site, the initial Web site, it can be pointed to from a second Web site and contextualized for its architect's own purposes regardless of where it was originally found and resides. The Web is modular, and the modular units that make it up can be used one or more times, often for purposes that were not originally envisioned.

Modular units can be combined and recombined to produce larger entities that are variable. "Variability" is Manovich's term for the characteristic of new media which means it is never "fixed," but exists in "potentially infinite versions" (36). The principle of variability is easy to see in hypertext technologies, as the links in or between Web sites allow the user to make an inexhaustible number of sequenced choices as she moves from one hyperlinked file document to another.

It is this principle of variability that has led many to believe that new media technologies celebrate individuality, as we all have our own choices about how we engage with the electronic documents we work our way through, what Manovich calls the "logic of new media" versus the "logic of industrial mass society," where everyone had to "enjoy the same goods" such as traditional print and cinema (41). However, some might challenge the idea that new media allows for more individuality than traditional media; even though we read a book and see a movie from start to finish, we still bring our own sensibilities to and create our own meanings from these traditional texts, and they too allow for freedom. Because of our unique life experiences, certain words or characters in books or movies might have different meanings for each of us.

Conventional forms of media such as print texts are based on narratives that are a series of scenes or dramatic moments captured in a specific order to convey the specific meaning of the author or director. The scenes by themselves can be read or viewed and enjoyed, but narratives are built on the specified order of scenes. In contrast, Manovich points out, databases possess many separate modular elements that can be variably combined and recombined; what comes before or after each element in a database is of no consequence, and each element is just as significant as any other element (218).

Rather than just extending the concepts of variability and modularity, Manovich's database is important to understand as a "cultural form" where "the world appears to us as an endless and unstructured collection of images, texts, and other data records" (219). Not all information has to be tied together to have value; a database seems to suggest something complete in itself, something that has a value without the narrative ordering found in traditional media. In contrast, traditional narratives are thought of as incomplete or senseless if they are unstructured; we like our plots presented to us in the right order and with enough information in each scene that we can follow and understand it.

The Internet is where we can really see this cultural form in place. A Web site consists of modular and variable texts, images, and data records that can be added to or trimmed down without losing the sense of purpose. Manovich makes the point that it is easy to add a line of HTML code to include more information to a Web site which is "a sequential list of separate elements" (220). When contrasted with other forms of traditional texts such as movies or novels, it is harder for us to imagine that we can just add or subtract information after they are complete; for everything to make sense in the narratives, we need to have all of the proper elements in place—the plotlines, backgrounds, and character development that would allow us to believe that the actors or people in the narrative would plausibly do what they do.

Take a look at any Web site and imagine removing something from it or adding to it. Would this action really change its sense of wholeness? Can you imagine all these separate elements as data as opposed to indispensable elements that we need to see in a narrative? Like databases, Web sites can always be added to or subtracted from but still retain a sense of wholeness. A novel with a couple of chapters missing cannot do this.

Perhaps you have seen Web sites that have very little information put up by the architect, but instead consist mostly of links to other Web sites. Go view the opening page of someone's social networking site such as Facebook and note just what elements make up the Web site. A person's "friends" or links to other sites that suggest something about the person? What does this suggest about who a person is? Can someone really define herself by using only links to other people and Web sites? How does this differ from a traditional text such as a personal written essay or an autobiography? Are social networking Web sites nothing more than databases? Have Web sites become a "cultural form" that is based on a database?

To describe our present day comfort with this database form of new media, Manovich borrows from Charles Baudelaire, a nineteenth century French poet, who wrote of the Parisian *flâneur*, a "new modern male subject" who strolled through the streets of Paris, noting the faces of people he passed, and immediately erasing them and moving on to others. A *flâneur*, which is French for "stroller," finds comfort in movement through a crowd:

> To the perfect spectator, the impassioned observer, it is an immense joy to make his domicile amongst numbers, amidst fluctuation and movement, amidst the fugitive and the infinite ….To be away from home, and yet to feel at home; to behold the world, to be in the midst of the world and yet to remain hidden from the world.
>
> Baudelaire (quoted in Manovich 269)

Moving through a crowd as an "impassioned observer" on a Parisian street, with all of its well-dressed and attractive people, while retaining the freedom of not having to make any lasting connections comprises the *flâneur*'s sense of self or subjectivity. The *flâneur* is perhaps compensating for his lack of personal relationships by mingling with the anonymous crowd in an exciting urban center, and this is the effect of the modern world on us. At most, the *flâneur*'s ability to connect with people is reduced to a "split-second virtual" moment where he notices and is noticed by another person, as they share a fleeting intersubjective connection, then end it and move on (Manovich 269).

Like Baudelaire's *flâneur*, who finds pleasure and fulfillment moving through a Parisian street, connecting with people for a nanosecond, then moving on, today's Internet surfers "travel through virtual streets, highways, and planes of data" and are always "happiest on the move" (Manovich 274). This would not be possible if we did not have the stored databases available on the Internet at the ready to move through. When you surf the Internet, in what way is your experience like that of Baudelaire's *flâneur*? How does surfing the Internet allow you to define yourself, give you your sense of subjectivity? How might this be different when you are spending time with friends or family? Are you "happiest on the move" and connecting for only a "nanosecond"?

Lev Manovich challenges the "myth" that modern communication technologies such as hypertext are in some way more interactive than traditional technologies. To be "interactive" really means more than pushing some buttons on a keyboard or choosing some hyperlinks to find your way through a Web site. To be truly interactive, a technology would have to reflect the specific and idiosyncratic ideas and emotions of each person who engages with a computer-based technology, and this has yet to be achieved.

This myth has been around for some time and can be seen in the tendency of psychological theorists such as Sigmund Freud to equate "mental processes" with technologies such as photography that externalize our thoughts. For example, it was believed that a person's abstract thought could be accurately represented as a grouping of separate photographic images that were taken and then arranged and printed out as one photograph (Manovich 59). The belief that a human mental process as complex as our imagination could be equated to external representations such as these photographic collages was held by a number of prominent psychologists. When we use contemporary graphics programs such as Photoshop to present an idea, can we really capture what exists in our imaginations? Do we take it for granted that technology, if used in a certain way, can accurately reflect our thinking? How do our ideas about graphics programs compare to what the early psychologists and other people in the same age used to believe about the advent of photography and its ability to capture and register our most complex thoughts?

Manovich also suggests that the belief that we can externalize complex human thought by using technology was based on the need of modern industrial society to standardize ideas (60). Our society is built on manufacturing methods that require the production of many identical parts which we then build things with. We have transferred this concept to our view of human mental processes, and because of this they are not presented as being as unique as they really are. Perhaps we can see this in images used in advertising, where the wants and desires of us all are reduced to one-size-fits-all imagery. Think about all of the things you are and how fluid your imagination is. Could you put fifteen images up on a social media site that really tell people who you are?

Thus, that the word "interactive" is so often used when we hear people discuss new computer technologies should not be a surprise to us—this tendency has been in place for a while. Regarding hyperlinks, Manovich asserts that when we see an image, word, or sentence in a book, we bring our own unique associations to it. Our reactions to movies and books can be tangential and imaginative and the associations we make are built on a "logic" that is unique to ourselves. However, when we select an image or some underlined text in a hyperlinked environment, we are sent off to other places that others have decided we should go to and which are "pre-programmed" (61). We think we are making unique choices that reflect who we are as individuals, but this is illusory.

Manovich paraphrases the philosopher Louis Althusser when he writes "we are asked to mistake the structure of somebody else's mind for our own" (61). That we have a database-like structure on the Internet that connects discrete or separate structures through links does not mean that we are in a truly interactive environment. Just how interactive is a Web site? Are you interrogating it and asking questions of it as you work your way through it? How might it be akin to a conversation or discussion you are having with someone? How does this differ from reading a book? If you are reading a book carefully, are you not challenging some of the assumptions of the writer, or are you just agreeing with everything that is said? When you are reading a book, do you see the letters and the words of the alphabet in a different way than you might on a screen where you can change the text size or screen size?

The age of print and the late age of print

Writing has had a profound effect on us, as Walter Ong describes, and in this section ideas about what writing is and its place on the computer screen will be characterized and contrasted with traditional writing.

A "writing space" is a "material and visual field" that we compose on and read from, and it can take the form of such technologies as papyrus, paper, or a computer screen. J. David Bolter is interested in how the changes in writing spaces have affected our ability to think and communicate with each other: "Writing, even writing on a computer screen, is a material practice, and it becomes difficult for a culture to decide where thinking ends and the materiality of writing begins, where the mind ends and where the writing space begins" (12–13).

Bolter's ideas about the effects of writing spaces on us parallel Ong's belief that "writing restructures consciousness" (Ong 77). Bolter advances Ong's project by characterizing two historical ages that in part lie beyond the orality and literacy periods. They are the "age of print" and the "late age of print." Printing replaced handwriting because so many more copies of a manuscript could be produced and then disseminated. Before the printing press, people copied manuscripts by hand, which took a lot of time and was costly.

The late age of print characterizes the age we are living in now, where we still use print but in electronic formats such as word processing programs, email, and the World Wide Web (Bolter 2). While we still use books, we also use print in digital format and at an ever increasing rate. We live and read in an age where the traditional print form is indispensable, as some information that we need can still only be found in print, but it no longer seems indispensable, as we are transferring more of our traditional texts to electronic format (Bolter 2). Whether digitized print will eliminate traditional print is still being debated. For many people, especially scholars, it is important to be able to get their words into print as this confers a higher status on their work. However, online journals that have no print backup counterparts are becoming accepted as the equals of traditional publications, and many scholarly journals, newspapers, and magazines are found in print and online.

Because of the permanence of traditional printed material, we somehow imagine these texts as unchanging monuments produced by authors who loomed above us. Great books and other texts were arrested in print and that was that; they were deemed perfect and thus should not be changed. This created respect for the author, but put some distance between the reader and the author. In the late age of print the impermanence of the written word undermines the sense of texts as monuments, since we know how they can be changed. We have all benefitted from the use of word processing and desktop publishing software and produced texts for our own needs, and because of this we are empowered as writers. However, Bolter notes that this flexibility "threatens the definitions of good writing and careful reading" that have come into existence during the age of print (4). Because we can write and change what we write so readily with word processing software, what is written loses its force as a powerful, perfect, and unchanging body of words.

Remediation

To better understand how the age of print and the late age of print differ, we need to understand what Bolter means by the term "remediation."

Remediation describes the shift to a newer form of media that takes some of the characteristics of a previous form, but then refashions it. Because the newer technology "remediates" the older one, there is an implicit assumption that the newer form improves it. For example, the process of producing books was remediated in the move from copying by hand to using the printing press; because books could then be mass produced, this represented an improvement (Bolter 23). The Internet that allows us to send email remediates the telegraph as a technology and as a "cultural symbol." Both technologies are made up of grids of electrical lines that cover the world and allow us to communicate with others instantaneously (Bolter and Grusin 197). When telegraphs first came into being, they symbolized to us that people could send instant written messages to each other, a remarkable improvement over the speed of nineteenth century postal services: imagine the time it would take for a letter to get from America to Europe, or from New York to San Francisco before the telegraph. However, today we can answer email from our home or our desks at work, whereas telegraph messages were sent to operators who had to decode the signal and then send it to the recipient by local courier. Also, telegrams were expensive, and email is not.

As a concept, remediation illustrates how all media are based on signs or texts that refer to things that are continually reshuffling and changing their relative values as we move on to newer communication technologies (Bolter and Grusin 19). When one kind of medium remediates another, the kinds of texts it uses are weighted differently and thus convey a different world to us. As people began to write things down on papyrus, when oral communication was remediated by written communication, people could still hear the words as they read them, but they could also see them, thus "giving the words a different claim to reality" (Bolter 23). They were weighted differently in part because we could both see and hear them. By seeing them, they moved beyond sound and had a visual equivalent or presence. The signs are reshuffled and revalued. When you read, do you hear the words? If this is the case, how might reading something have a different effect on us than hearing something? The signs in this case are the sounds of the spoken word and the written word combined. How has their value changed?

Imagine how email remediated other forms of communication such as traditional mail. In a way, it privileged a kind of writing that was not as formal as the kinds of letters that people used to write to one another. Penmanship is an important element in the epistolary or letter writing form of communication, but when people use keyboards to tap out the words in email messages to each other, good penmanship does not matter. The words used in letter writing convey things differently just by the way they look; some believe that a person's character can be gauged by the quality of his or her penmanship. When email remediated traditional mail, the texts used took on a new form. Emoticons are not part of the regular vocabulary that letter writers use, and email messages, for many, assumed a slapdash kind of writing where challenged grammar and punctuation practices such as starting sentences without a capital letter were not seen as detrimental to meaning or form. Can an emoticon present the precise meaning that a carefully handwritten sentence conveys? How many subtleties can be evoked with the same happy face emoticon? Can one be happy and concerned at the same time, mindful that the circumstances that led to the joy we are feeling might change? There is no emoticon for this kind of happy, a kind of pensive happiness. In a way, emoticons might be like the objects in Manovich's database culture, as we can insert these shorthand substitutes for complex feelings into our emails to produce a sort of modular communication.

Perhaps these differences are a function of the speed at which we write email; someone could write a message to you, you could quickly respond to it, then the person could respond to your reply, all within the space of a few minutes. Perhaps speed undermines formality. When we wrote a letter to someone we were perhaps more careful to get it right, as we had to put it in the mail and perhaps wait a week for a response. What we wrote could not be quickly corrected if the recipient got the wrong idea, so people made sure that what they composed would accurately convey what they meant. Of course, this does not mean that some people do not write elegant email messages and that some people did not write letters that were quickly written and not thought through very carefully.

I have suggested that a medium that has remediated another medium conveys a different world, that technologies are not neutral. Might the way we communicate to each other and thus relate to each other as human beings have changed with the advent of email and the subsequent decline in the use of the traditional letter that we send through regular mail? Perhaps you could make a case that email is still a better form of communication and that it more accurately reflects what we are really thinking. What is it about this medium that might make this true? How might the way we relate to each other via email affect us? Do men and women have different relationships because of email? Is exchanging email more like a conversation then sending and receiving traditional letters?

Transparency erases all sense of the medium that is presenting information to us. When things are transparent, we forget that we are looking at something that is really representing something else. This is similar to Richard Lanham's idea of letters as seen through rather than seen. We do not see the texts for what they are, as things in and of themselves. We think they equal the thing they are representing. When a technology becomes transparent, we do not see it standing between us and the message we are trying to send; it is as if the medium has no effect on the message being transmitted. For example, when reading a book, we forget that we are reading symbols called letters that make up our alphabet, and that make words, sentences, and paragraphs. The world is presented through language, which is a technology, and it is not the real world. It serves to represent the world. Landscapes and portraits in what are called "realistic" paintings try to get us to forget that the essence of the land or the people being painted are really just brushstrokes on canvas, but we do not see them (Bolter 25). When we look at landscapes and portraits online, we do not see the pixels.

Sometimes people who work in a particular medium want us to know that we are actually looking at the medium; that we should look not through it but at it; that it is not transparent. The television show *Saturday Night Live* will often show studio professionals, who are usually concealed behind the cameras, and actors scurrying into place before the next segment to remind people that what they are doing is in fact live television. Contemporary situation comedies such as *The Office* or *Modern Family* sometimes have the actors talking directly to the camera, as if they are part of a documentary, highlighting the use of such media techniques to manipulate our feelings or responses. In this case, they are applied to a staged comic effort as opposed to a sober and serious documentary, with the effect of making it seem like we know the characters and they trust us, or making the characters even more comic and absurd, as they are using the medium in a clumsy way. It makes the medium opaque, not transparent. It makes us realize that this is all staged, all made up.

"Hypermediacy" describes media that constantly remind us that we are looking at a construction—the opposite of transparency—and "offer themselves as immediate experiences" (Bolter 25–26). Some Web sites seem to celebrate the fact that they are Web sites: "In the logic of hypermediacy, the artist (or multimedia programmer or web designer) strives to make the viewer acknowledge the medium as a medium and to delight in that acknowledgement" (Bolter and Grusin 42). We are made conscious of this on some Web sites because of their "windowed" spaces, with many icons and links that "tr[y] to reproduce the rich sensorium of human experience" for us, presenting a seemingly unending array of choices of written, audio, and visual representations of the subject matter (Bolter and Grusin 34).

When you are online and viewing a Web site and enjoying the experience, is it because you are looking through the Web site and engaged with the material it presents, or are you cognizant of and enjoying the Web site itself—the choices you can make with the links, the manner in which it organizes information, and its basic look or aesthetic? Is the Web site transparent, or is it presenting itself as a hypermediated space? Are you in fact marveling at the ingenuity of the Web site, or the actual content that it is trying to represent?

The logic of hypermedia is reflected in the fact that Web sites are often designed with so much information in them, with images "crowded" together and many choices presented at one time (Bolter and Grusin 269). While some Web sites have remediated some visual aspects of television, television news shows have in turn remediated Web sites with the many choices of information presented on them. For example, news outlets like CNN will have an announcer presenting the news of the day in the center of the screen, but also horizontal bars above and below the newscast that replicate those first seen in framesets on Web sites, showing scrolling information alerting us to stock quotes, changes in the weather, or even more late breaking news than the announcer can tell us about from the script she is reading. It appears that television news has to cover even more if it is to compete with news found on the Internet, so the look of television news remediates a busy Web site.

The advent of digital technology has been a more traumatic remediation of the writing space than others in the past (Bolter 24). When the move was made from hand-copied manuscripts to printed books, the books that were being printed still looked like the previous technology. They resembled careful reconstructions done by hand, such as the Bibles that were painstakingly copied by monks in monasteries. Chartier maintains that word processing technology "changes methods of organization, structure, consultation, and even appearance of the written word" when contrasted with books (quoted in Bolter 24). Programs like Microsoft Word make it so easy to make written texts look like anything we want that we have moved away from the previous and rather unadorned look of books.

While the electronic writing space that Bolter speaks of is a traumatic remediation of previous writing spaces, electronic media in general have not been designed in "opposition" to photography, painting, and pre-existing texts (45). In many ways, electronic media have been developed to be transparent and to allow us to view older works of art with greater ease without challenging the authority of these art forms. For example, we now have digital access to the collections of the Louvre in Paris or the Metropolitan Museum of Art in New York, and we can see digitized representations of traditional art forms presented as faithfully as they can be, as if we were standing before them in the bricks and mortar museums.

It is for this reason that remediation is a defining characteristic of electronic media (Bolter 45). In a very short period of time applications of new media technologies have been designed to present all the traditional methods of conveying texts, whether letters, writing, art, music, or other recordings. We have been developing media forms for thousands of years, and in just a few short decades we have been hurrying to remediate them in digitized form.

Bolter and Grusin point out that the importance of the Mars landing of 1997 was that "this medium could itself provide an authentic and exciting viewing experience" (268). What was happening was that a small machine called a "rover" was slowly traveling across the landscape of Mars and capturing images of the terrain of the planet via an onboard camera. The still pictures were posted on the Jet Propulsion Laboratory's Web site as they came in, so we could all see them and not have to wait until they were on the evening news. Perhaps what was missed was the fact that humans had safely landed a machine on a planet in our solar system and we could see what it actually looked like for the first time in our history. While none of the humans who had lived before us had ever seen what Mars looked like, the event was presented in a way that made the technology that allowed us to see it more important than the images themselves. When a technology conveys the first viewing of Mars by human eyes in this way, the medium is more important than the message.

When you read about a new information technology, how often do the writers seem to stress the breakthrough in the engineering rather than the fact that the technology allows us to do something important such as improve education, medicine, or the environment? Is it about the technology, or about how the technology can help us? Perhaps you might review an article in *Wired* magazine and identify what is stressed.

Words on pages and screens

Sven Birkerts is not someone who thinks highly of the transfer of words from what has been the most common text, the book, to the computer screen, and his reasoning is compelling. A word in print on a page, while "nearly weightless," has some substance and is "verifiably there," whereas a word on a screen is different in that once the screen is turned off, it is "dematerialized, digitized back into storage, into memory, [and] cannot be said to exist in the same way. It has potential, not actual, locus," or location. Birkerts moves from a technical to a natural analogy to emphasize this point when he writes that the word on a computer screen "floats on the surface like a leaf on a river" and is "less absolute" than "the leaf plucked out and held in the hand" (155–156). For Birkerts, if you close a book and put it aside, the words in ink will still be there, albeit in the dark and out of sight, but when you turn off a computer, they are gone.

We can extend this insight; the medium that we read texts in influences their power and authority:

> Wouldn't we say that the word cannot really exist outside the perception and translation by its reader? If this is the case, then the mode of transmission cannot be disregarded. The word cut into stone carries the implicit weight of the carver's intention; it is decoded into sense under the aspect of its imperishability. It has weight, grandeur—it vies with time. The same word, when it appears on the screen, must be received with a sense of its weightlessness—the weightlessness of its presentation. The same sign, but not the same.
>
> Birkerts (155)

The medium we read the word on—whether it is paper, screen, or rock—has something to do with how we perceive it. If a stone carver took so much time to carefully chisel out some words in stone, we feel that these words are more important than words that have been dashed off and instantly show up on a computer screen because the stone carver puts in so much focused physical and mental effort. The writer using a pen or typewriter is "more likely to test the phrasing on the ear, to edit mentally before committing to paper" because it is just harder to write and/or type something out and then have to rewrite or retype it later (Birkerts 157). The traditional writer thinks carefully about what she is writing, goes through all the word choices and phrasings in her mind, not on the screen, and then puts them on paper.

However, perhaps Birkerts' ideal writer, who carefully thinks through every word choice, could do the same on a computer; the mental process of this writer is just facilitated with greater ease—instead of trying to conjure up the correct words in our heads, we do it on the screen. When we get it right, we stop and save it. Then we print it. Do the words on the page that eventually come from this process have any less meaning, power, or gravitas than words on paper or stone? Imagine reading the above passage by Birkerts after it had been carved in stone. Would you attribute a different meaning to it? What about if you read it from an ebook on a computer screen?

The act of writing is more inviting when using the personal computer, and involves a tool that more accurately reflects the process that good writers engage in when they write. Writing on a computer makes it easy to experiment and rewrite, and this is a healthy way to think about writing. However, Birkerts points out that writing using a computer, which in some ways represents "the very apotheosis of applied rationality" because of the complex technologies involved in its design, is, ironically, "destabilizing the authority of the written word and returning us … to the process orientation that characterized oral speech" (156). As Ong has pointed out, orality is more dynamic, and Birkerts feels that the "unstable" way of producing information on a computer is like oral speech, where we work out our ideas as we talk or argue. This diminishes the power of the written word produced by the writing practice where one carefully figures out each point or idea, writes or types it out, then moves on to the next idea. By "process," Birkerts means that writing on computers has us understand that words are provisional and we write them out more quickly, then change them without thinking anything of it (157). While it is easier to write on a computer, this practice undermines the stability of the writing we produce; words that we produced by traditional means can be "transmitted, studied, and annotated" because they were "reared on a stable base" without the sense that, unlike words on a screen, they would soon be deleted, replaced, or altered (Birkerts 156).

As we move away from book culture to a culture based on electronic communication, Birkerts argues, the quality of the prose we use will diminish significantly in a process he calls "language erosion." As opposed to using language that is unique and nuanced, we will have to live with what he calls "plainspeak," where "Simple linguistic prefab is now the norm, while ambiguity, paradox, irony, subtlety, and wit are fast disappearing. In their place, the simple 'vision thing' and myriad other 'things'" (129). Stock phrases are certainly part of our culture and it is easy to add the word "thing" onto a word to explain our feelings about something in a "prefab" style of speech, or perhaps "prefab thing," but is this really something that exists *because* of email, text messaging, and the Internet? There have always been sayings in every age that people have used as verbal shorthand to express their feelings about their experiences; this did not start with text messaging or email.

Even if this is the case, this does not mean we cannot do better, and Birkerts shows us how when he writes. He laments the loss of the habit of reading great literary works of writers such as Virginia Woolf and James Joyce, as this enhances our ability to use language for the benefit of our culture: "the civilizing energies of their prose will circulate aimlessly between closed covers" (129). This quote illustrates that, as Birkerts writes, he does so in a facile and meaningful way to accurately reflect his concerns about the loss of great writing. The image that he leaves us with in this quote is of the covers of an unopened book containing great ideas that "circulate aimlessly." He characterizes these ideas as animate entities; they are alive and trapped inside books only to wander alone, unused and ignored, unexploited intellectual resources at a time when, like all times, we need ideas. If Birkerts had written "It is a no-brainer that we should read literature," he would not be telling us why this is important or presenting this idea in a compelling way. To use the stock term "no-brainer" is too easy. To say that if we do not read literature "the civilizing energies of their prose will circulate aimlessly between closed covers" is more thoughtful and challenges us to think about what is really going on between the covers of a great book.

When one relies on retrieving information from a screen as opposed to going to a library, our sense of "historical perspective" is flattened (Birkerts 129). According to Birkerts, traditional libraries, with their shelves of books aligned in an organized fashion, present a visual and concrete version of the depth of history. When one goes into a library and walks around the stacks, it is easy to see the ages of history represented spatially: "we form a picture of time past as a growing deposit of sediment; we capture a sense of its depth and dimensionality" (Birkerts 129). We can imagine different historical ages because the books that are devoted to each one, whether ancient times or the time of the American Civil War, can all be found in the same place, separate from other ages. If you looked down a long row of books on history you could see the different ages represented relative to one another, as if they were a timeline; just walking along a row of books gives you a sense of history.

When we use a personal computer to retrieve historical information, we are presented with search engines that do not provide this historical context. As described by Manovich, databases in general present information in a modularized fashion, as "a body of disjunct data," and we do not get a sense for the context of the data. Even a page in a book gives a sense of how the information it gives is to be contextualized, just because of where it is relative to the other pages in the book. Because it is in a book, it implies that it is one important part of a larger and more complex picture that we need to understand; it is one part of the whole story. Without this perspective, we will become more "rooted in the consciousness of the now" and "it will seem utterly extraordinary that things were ever any different" (Birkerts 129). Recognizing the differences between things is the key to understanding them, and a failure of perspective would diminish our ability to have a clear and robust understanding of history. If everything can be plucked from a database and found on a screen, everything can be understood to have the same value.

While Birkerts describes these flattening effects in terms of recorded history, we could expand this idea to everything in libraries, as they are organized in a way that separates and compartmentalizes knowledge; in one section there are books on physics, in another section there are books on government. These two areas of study do not intersect very often and that they are in different parts of the library allows us to better comprehend their differences. Physics is over there, next to chemistry; those fields deal with math and atoms. Government deals with words and law and people trying to work out their differences; it is closer to the books on history.

Birkerts' most ominous suggestion about electronic media is that it contributes to the "waning of the private self" (130). While in part this means that we will lose some privacy, it also means that we will to some degree lose our sense of identity and merge with everyone else in our society. As the "monitor light is always blinking … and we are always potentially online," we run the risk of not being able to keep our distance from the electronic systems that connect us all. While we are connected via the screen, our lives become more available and accessible, and people know more about us. If we are always connected, if everyone knows where we are, what we think, and what our likes and dislikes are, where is the quiet and distance we need to separate ourselves from others?

Perhaps we have been made to feel that the contrary is true. For example, one of the most common claims for social networking sites is that you can indeed stake your claim on the Internet and let everyone know who you are as an individual, but in many ways these sites are based on cookie-cutter templates where we are asked to tell people who our favorite musicians are and who we admire. This can tell others something about us, but who we are might be more difficult to convey than just by setting up a series of links to other organizations and people. That's not who we are. How can we be unique if we are constantly interacting with others who deluge us with their thoughts and also know ours? Can an individual live outside the constant engagement in online communities? Are we always mindful about what others think of us because we are always on display for people to examine?

Manufacturers of personal computers and their attendant software have exploited our felt need to be individuals when we use the machines they are selling us, but can this really happen? In many ways, they are asking us to join with them and the community of those who also use the same machine, but do we need something we own that connects us to others to characterize who we are, or is this just a company's way of selling us something? When we are told that "There's power behind every nanometer" on the new MacBook Pro (Apple), does that really make us powerful if we buy one?

Birkerts describes his version of the figure–ground model where an individual stands before the backdrop of all others in our society and we can clearly distinguish the separation. Imagine yourself, standing in front of a crowd, everyone else in the world, and in some way, your distinct character comes out as an "exalted individual." Is this possible if your personal computer is "always blinking" and you run to it? Engaging with others is an important part of being a complete person, but in Birkerts' view we get too much of this and become so entangled with everyone else in online environments that we cannot really distinguish ourselves and be unique.

Standing alone, quietly, away from frenetic interactions with others on the World Wide Web better enables us to figure out who we are and be unique.

McLuhan and media's messages

Marshall McLuhan's "the medium is the message" is a phrase many of us have heard before, but perhaps some are not quite sure how it should be applied. When McLuhan uses the phrase he means that to really understand the medium—the technology that we create—we need to understand its larger effects on our culture or society. These changes are the message.

When a new communication technology comes along many people celebrate it by forecasting how we are going to be able to move information around faster and receive it in more locations, but according to McLuhan, these observations miss something more important. What we fail to notice, or perhaps do not notice at the time of a new technology's inception, is that eventually it has other effects on our culture, and this is the "message" that McLuhan refers to: "the 'message' of any medium or technology is the change of scale or pace or pattern that it introduces into human affairs" (8). The message, according to McLuhan, is not the information that is presented with the new medium; it is how the technology itself changes our culture.

For example, McLuhan tells us that a light bulb has no content at all; it just emits light. However, electric lighting has changed the way we live and he cites night baseball as something that would not exist if we did not have this technology. We can extend this to other areas. The "message" of electric light is that we can work inside and at night if we desire, without ever imagining that this would be difficult. In the past when we did not have artificial light we never had any idea that we could do this. Artificial light has allowed us to imagine where we can work and live.

We can apply this concept to cell phones. One of the joys of cell phones is that people can always make contact with us, but this is in some ways also a drawback, because we can never get away from people; if we give someone our cell phone number, there is an implicit "message" that we are always available and this can cut into our privacy. Also, because we get used to always being able to speak to someone wherever we are, we do this instead of buckling down and doing our work, whatever it may be, just being content with the idea that we can spend the five or ten-minute period that it takes to walk to lunch, another class, or another appointment alone, and enjoy the sunshine and the quiet of the day. So it is not the message sent over the airwaves and received via a cell phone that McLuhan is writing about. Instead, it is the message that we have created a new level of connectedness with other people, that we are always connected, but that we have lost some of our ability to be content with our own thoughts or focused on other things.

Nicholas Carr's criticism of the way we use the Internet in many ways takes the approach Marshall McLuhan might have applied. Carr points out that the Internet is a "godsend" to him, as it has made the process of research more convenient and he does not have to spend "days in the stacks or periodical rooms of libraries" (2). However, he also relates anecdotal evidence from friends who spend increasing amounts of time reading snippets from the constellation of Web sites on the Internet that suggests that this practice has diminished their ability to concentrate and read and think about longer pieces of information. The technology allows us easy access to information, but this technology has also changed our relationship to it, as now we do not read as deeply as we used to; instead, we skip from URL to URL.

Carr characterizes the belief of the professionals who invented and maintain the Google search engine—that our ability to find information quickly allows us to be better thinkers—as a simplistic reduction. This is to some extent a fine idea, and for a long time we have had libraries that served this purpose, but Carr's concern is based on the increased ease and efficiency that Google provides, which allows us to believe that the insights we need for real problem solving can come to us easily if we just have some facts to hand:

> It suggests that intelligence is the output of a mechanical process, a series of discrete steps that can be isolated, measured, and optimized. In Google's world, the world we enter when we go online, there's little place for fuzziness or contemplation. Ambiguity is not an opening for insight but a bug to be fixed. The human brain is just an outdated computer that needs a faster processor and a bigger hard drive.
>
> (173)

There is a difference between possessing facts and using them to gain larger insights or knowledge, however we see the belief Carr describes running through other parts of our culture. On the long running television show *Jeopardy*, seemingly well-read contestants are rewarded for responding with a question such as "Who is John Wilkes-Booth?" to a prompt such as "Stage actor who assassinated Abraham Lincoln." It has often been believed that these skilled players on *Jeopardy* are highly intelligent, as they can marshal their answers so quickly; it is as if they know everything. But is this really intelligence? Regarding the word "ambiguity" in Carr's passage above, does the everyday use of Google allow us to believe that there is always an answer to difficult problems that we can find with a fact supplied by Google or some other body of information found online? What would Marshall McLuhan say about this technology and the way it might have changed not what we think, but the way we think? Does it imply something like "All you have to do is go online, find the right Web site, and it will provide 'the answer'"?

When Carr suggests that perhaps ambiguity should sometimes be welcomed as "an opening for an insight," we might imagine a difficult problem we have had in our life that we mulled over for days or longer. While this process is not always easy and at times frustrating, perhaps the things we have been compelled to think about as we work our way through this process made us realize something that we could never have understood had we just had a few facts to contemplate. Finding a few facts and then making a decision is a method for solving simple problems.

Recently, the professionals at IBM were able to program a computer they named Watson so it could actually "beat" the human contestants on *Jeopardy*. Are we to assume that Watson is more intelligent than a human? If Watson was asked to do something such as convince a state legislator in your district to reverse their decision to vote a certain way on a major bill, would Watson be able to do so better than a well-informed and rhetorically able citizen who has been studying the issue? Could Watson change a politician's mind better than a human? Perhaps being able to find, arrange, judge, hold on to or discard, and then use facts is different than just knowing them. Being able to decide what is important sometimes requires us to look into the enigmatic face of ambiguity.

Similarly, American and NATO forces in Afghanistan have developed an analytic culture that to some extent is based on presentation software that relies too much on disparate or unconnected facts. What worries some of the officers is that those who base their wartime briefings on PowerPoint slides compel too many decision makers in the military to believe that all one has to do to explain a complex and dynamic military problem is produce a set of unconnected concepts or a list of facts or points—this sounds like an insight about the medium controlling the message that McLuhan would support. Like Google and *Jeopardy*, technologies that grant us easy access to facts or concepts do not necessarily help us solve real and difficult problems.

Some commanders in the field understand the weaknesses of presentation software and have pointed out that the challenge of writing a five-page paper demands that an officer work harder to present "an analytic, persuasive point" than the software, which "relieves" officers of this burden. One American general characterizes this practice as follows: "It's dangerous because it can create the illusion of understanding and the illusion of control" and "Some problems in the world are not bullet-izable" (Bumiller). Texts that tend to distill concepts such as "Population Conditions and Beliefs" and "Coalition Execution Capacity and Priorities" have value, but when presented as a series of separate points which are not explained in depth or tied together in a traditional written format, with nuanced ideas, background information, and connecting transitions, they do little to really help decision makers. The idea that the cultural and religious beliefs of the population of Afghanistan can be reduced to a pithy phrase by western Europeans and Americans, when previous foreign interventions in the country over the last several centuries have failed to reach any comparable level of understanding, is asking a lot, and the use of technology could mask this insight. Presentation software does not necessarily allow us to entertain the ambiguity that Carr mentions is often present in a-real world situation, and which invites more reflective thought.

Technologies such as Google have changed our relationship not just to facts and information, but to the process of reading. Carr enlists the work of Maryanne Wolf, a developmental psychologist, to point out that when children learn to read today, they work more as "decoders" than deep readers (Carr 122). Wolf calls for a more involved style of reading:

> I do wonder whether typical young readers view the analysis of text and the search for deeper levels of meaning as more and more anachronistic because they are accustomed to the immediacy and seeming comprehensiveness of the on-screen information—all of which is available without critical effort and without any apparent need to go beyond the information provided. I ask, therefore, whether our children are learning the heart of the reading process: go beyond the text.
>
> (225)

The "seeming comprehensiveness of the on-screen information" that Wolf is concerned about is comparable to online information that does not invite the "contemplation" that Carr draws our attention to.

Wolf's conclusion that we need to "go beyond the text" is to some degree extended by the work of Ian Rowlands and colleagues in the article "The Google Generation: The Information Behaviour of the Researcher of the Future," but some important things are also added. An assumption of many middle-aged and older academics is that the young scholars who make up the "Google generation" are different in that they have been distinctly shaped by search engine technology. However, an examination of data on the computer search habits of scholars across generations shows that they all share something in common:

> CIBER deep log studies show that, from undergraduates to professors, people exhibit a strong tendency towards shallow, horizontal "flicking" behaviour in digital libraries. Power browsing and viewing appear to be the norm for all. The popularity of abstracts among older researchers gives the game away. Society is dumbing down.
>
> Rowlands et al. (300)

It is a "myth" that only younger scholars read in a shallow manner (Rowlands et al., 300). Online consumers of information across generations have a different relationship to information, one which McLuhan would see as a function of the technology available to them.

Think about the "message" of other contemporary media in the way McLuhan would. For example, if you have taken an online course, imagine the effect that online teaching has had on our culture. How has it changed your concept of what teaching and learning is? How has it changed your thinking about teachers, students, colleagues, family, professors, or educational institutions? What might be the "message" of this technology? Is it changing the way people perceive universities? How has it changed our understanding of what a university classroom is? This goes beyond the content of the course you take, which is not what McLuhan means by "message." Think about other media such as the World Wide Web. What is the effect or "message" that they have for our culture?

Exercise

Go to the Appendix and review the edited letter that Sullivan Ballou wrote to his wife Sarah shortly before his death on the battlefield during the Civil War. Could you go through this letter and insert emoticons for some of his sentiments, or are his ideas so nuanced and unique that there are no substitutes for what he writes? Point out specifics. Perhaps Ballou uses some expressions that we have not heard before, but they were clichés when he used them. Can you identify them? Can you identify any thoughts that he has that seem fresh, that are not clichés but thoughts you can imagine Ballou came up with on his own to express his feelings to Sarah? Imagine you were able to see a graceful and dignified handwritten version of this letter. Would it have a different effect on us as we read it because of Ballou's handwriting? When contrasting the handwritten form versus the online version formatted in a conventional font, how are the texts or signs different? What does this tell us about texts that are remediated, even if they present the same content, the same wording? How might they present a different reality or worldview?

Works cited

Apple Inc. "MacBook Pro." June 25, 2012. Web.
Birkerts, Sven. *The Gutenberg Elegies: The Fate of Reading in an Electronic Age*. Boston: Faber and Faber, 1994. Print.
Bolter, Jay David. *Writing Space: Computers, Hypertext, and the Remediation of Print*. 2nd ed. Mahwah, NJ: Erlbaum, 2001. Print.
Bolter, Jay David and Richard Grusin. *Remediation: Understanding New Media*. Cambridge, MA: MIT Press, 1999. Print.
Bumiller, Elisabeth. "We Have Met the Enemy and He Is PowerPoint." *New York Times*, April 26, 2010. Web. December 12, 2012.
Carr, Nicholas. *The Shallows: What the Internet is Doing to Our Brains*. New York: W.W. Norton, 2010. Print.
Lanham, Richard. *The Electronic Word: Democracy, Technology, and the Arts*. Chicago: University of Chicago Press, 1993. Print.

McLuhan, Marshall. *Understanding Media: The Extensions of Man.* Cambridge: MIT Press, 1994. Print.

Manovich, Lev. *The Language of New Media.* Cambridge, MA: MIT Press, 2002. Print.

Ong, Walter. *Orality and Literacy: The Technologizing of the Word.* London: Routledge, 2002. Print.

Plato. "Phaedrus." In *The Rhetorical Tradition: Readings from Classical Times to the Present.* Eds. Patricia Bizzell and Bruce Herzberg. Boston: St. Martin's, 1990, 133–143. Print.

Rowlands, I., D. Nicholas, P. Williams, P. Huntington, M. Fieldhouse, B. Gunter, R. Withey, H.R. Jamali, T. Dobrowolski, and C. Tenopir. "The Google Generation: The Information Behaviour of the Researcher of the Future." *Aslib Proceedings,* 60.4 (2008): 290–310. Print.

Wolf, Maryanne. *Proust and the Squid: The Source and Science of the Reading Brain.* New York: HarperCollins, 2007. Print.

Chapter 2

The Internet and HTML

Chapter overview

In this chapter we will learn the basic building blocks of HTML and associated technologies that we need to begin formatting, saving, manipulating, validating, and linking written texts and images so we can produce Web site files. We will also use HTML for some simple layout techniques. This will be presented in the context of the history of the Internet, the work of Tim Berners-Lee, and the role of the World Wide Web Consortium, or W3C.

The beginnings of HTML

HTML stands for "hypertext markup language." While many of us know this, some historical context will better allow us to appreciate it. The word "markup" has been around for some time. Print editors used tags to "mark up" texts such as newspapers and novels to specify to the typesetter just where new paragraphs should start, what the text style and size should be, when there should be a line break, and other elements of layout and typesetting. Thus they used specific markup to indicate, for example, the beginning and the end of a paragraph long before HTML came into being.

When an editor received a manuscript that was to go to press, they would mark it up so that the printer, who had to set the type with his hands, would know how to make the text look. They did not type in tags around text. Instead, they took a finished paper manuscript, perhaps one that had been typed out on a typewriter, and then made their marks on the pages, using a blue pencil to distinguish the marks from the black type of the typewriter.

Of course, HTML markup tags such as <a href>, which is used to describe where a link should go, were not used by traditional editorial professionals as, other than a footnote or endnote, there are no links between words or phrases in books as there are links between hypertext documents.

Tim Berners-Lee simplified the rules of SGML, an older markup language, to produce the first version of HTML, and he is clearly identified as the person who put enough technology in place at the right time to allow the World Wide Web to come into being. While others spurred the growth of the Internet, it started with Berners-Lee.

Many people have good ideas that are never accepted because there is no community of people to begin applying an idea and making it the standard way of doing business. At the time of his creation of HTML, Berners-Lee was working at CERN, a particle physics laboratory in Switzerland, and he saw his invention as a tool for scientists to exchange ideas and to support advanced research. Berners-Lee's idea took hold because the people to whom HTML was first introduced were technically able and saw that they could actually use it; they understood, as Berners-Lee tells it, that "much of the crucial information existed only in people's heads" (Berners-Lee and Fischetti 9). This community of people saw the value of exchanging information to advance scientific discovery.

Berners-Lee makes it clear that there was not a "Eureka moment" when a vision of the Internet of today came into his mind. The process was one of "accretion" over time, with the invention of essential technical elements such as HTML that were actually used by people, and all these developments were based on the idea that "there was a power in arranging ideas in an unconstrained, weblike way" (Berners-Lee and Fischetti 3). Berners-Lee means that information can and needs to be compiled in ways that makes sense to humans, not necessarily to computers or machines, which are great at storing and arranging things in hierarchical fashion but poor at associative thinking: "the human mind has the special ability to link random bits of data. When I smell coffee, strong and stale, I may find myself again in a small room over a corner coffeehouse in Oxford; my brain makes a link, and instantly transports me there" (Berners-Lee and Fischetti 3). If a computer could smell coffee, could it possibly connect this sense to a "small room" that years earlier had a special meaning to one human being?

One of the virtues of the technologies that Berners-Lee developed was that all one had to do to be able to share information with others was upload it to a server on the Web, and browsers would translate it. To do this, and in addition to HTML, Berners-Lee needed to invent the following technologies:

> **URIs**—"Uniform resource identifiers" or URIs, which we now call "uniform resource locators" or URLs, are the addresses for Web sites that allow us to find a site on a server and review it without having to write complex computer programs to find and access the information on other computers (Berners-Lee and Fischetti 60).
>
> **Browser**—A browser decodes a Web site's HTML and allows us to read it. The one Berners-Lee invented was called WorldWideWeb (Berners-Lee and Fischetti 29). Today the most common browsers are Internet Explorer and Mozilla Firefox, and there is also Apple's Safari and Google's Chrome.
>
> **Web server**—This is the software that allows us to store information on a computer and allows others to access it (Berners-Lee and Fischetti 29).

The first server was on Berners-Lee's personal computer and was registered as info.cern.ch.

HTTP—"Protocols" are basic rules that allow computers to exchange information. "Hypertext Transmission Protocol" (HTTP) is a simple protocol language that allows Web browsers to find and identify the URIs or URLs of HTML-based Web sites on servers and translate the HTML so human beings can view and understand it (Berners-Lee and Fischetti 29).

With these URIs in place, the HTTP to help transfer the information from the server to an individual's browser on their computer, and the HTML used to format the text and other information on the sites, everyone could post and share their ideas and data.

In an age when people design, develop, and patent new technologies only to make money, it should be noted that Berners-Lee did not do this. Berners-Lee believes that patents undercut the development of the Web and its ability to serve the "common good" (Berners-Lee and Fischetti 196–197). Another problem he sees with patents is that it is difficult to identify when a new technology is "novel"; sometimes people would write computer code for a process or business practice that had been in existence for some time, then apply for a patent so they could receive economic rewards (Berners-Lee and Fischetti 196). For example, Jeff Bezos of Amazon.com sued Barnes and Noble because they began using a version of the "one click" order button that had been on Amazon's site. Because of this lawsuit, Barnes and Noble had to put a "superfluous" extra button on their site that said, after a patron had selected a book to purchase and given their payment information, "Please be sure to click this button. If you don't, we won't get your order!" (Gleick). What Berners-Lee wanted was for people to use these Web technologies he developed and then contribute to them so people would be able to connect to one another.

Berners-Lee does give significant credit to Robert Cailliau, another scientist at CERN, who understood what the World Wide Web could be, rewrote a proposal for it, and helped proselytize its development and use at CERN.

Berners-Lee originally thought that electronic editors would be used to code the HTML for text and graphics, but he discovered that "the human readability of HTML was an unexpected boon," and that people could learn how to use the HTML tags to produce documents on their own (Berners-Lee and Fischetti 90). What he had to convince people of was that, instead of navigation through a library system, even a library that was archived electronically, they had to go into a "space" where they would navigate through a number of Web sites that were connected to each other. This abstraction, which we take for granted now, became easier for the early users of the Internet and HTML to understand with the advent of bookmarks that allowed them to mark a page or pages using their URIs, then come back to them later, which gave them a sense of "persistence, of an ongoing existence to each page." Bookmarks also allowed people to use the "mental machinery they naturally have for remembering places and routes" (Berners-Lee and Fischetti 37). To Berners-Lee, the HTML allowed people to make links from Web sites and bookmark other Web sites in a way that replicated the natural associative patterns of their brains.

Having the technology in place and introducing it to people who would use it to search for information was one thing, but getting people to put information online was equally important to Berners-Lee: "The physicist would not find much on quarks, nor the art student on van Gogh, if many people and organizations did not make their information available in the first place" (Berners-Lee and Fischetti 38). The Internet would never have grown if everyone just consumed information and did not produce it. For this to happen, there could be no central control of the World Wide Web, so that anyone who wanted to put information online could. Imagine if there was only one server or one Internet browser, controlled by a person or group of people who decided what constituted the Web and who had access to it.

Beyond enabling people to contribute information to the World Wide Web, new technologies would have to be developed that would allow it to expand, and Berners-Lee hoped that more people would have a voice in deciding what new HTML and other related technological standards would be. To this end, he started the World Wide Web Consortium, also known as W3C, and this is still the most important organization for unifying the many parties that have a stake in the development of the Web.

While Berners-Lee invented HTML, HTML and other associated protocol languages are not owned by anyone. They are standards that W3C tries to get professionals from corporations and non-profit organizations to adopt for the technologies they will eventually design and manufacture, and they need to be updated from time to time. For example, browsers only work if they use HTML as this is the basis of the Web, the lingua franca of Web sites, and when there is a problem with some aspects of HTML the people who manufacture browsers need to suggest new ideas, perhaps propose new HTML tags that replace older ones that are ineffective, agree to compromise on some standards, and then actually use these standards so the WWW works smoothly.

Berners-Lee uses the analogy of the international phone system to explain his reasons for standardization of the basic WWW infrastructure: "The phone system defines what it has to, but then leaves how it is used up to the devices. That's what we needed for computers on the Web" (Berners-Lee and Fischetti 99). The signals and voltages for phones are exactly the same everywhere, and with appropriate adapters, different companies can produce different end-of-the-line devices, such as phones and answering machines, that are all connected to the international phone system infrastructure.

In the same way, technologies such as personal computers made by different manufacturers can be used to connect to and work on the Internet.

Berners-Lee refers to this as the "separation of layers," and holds that it is fundamental that these layers stay separate. The primary separation would be between the Internet and the Web. The Internet consists of the hardware and software that allows for packets of information to be separated and recombined by routers and then sent along their way over electronic networks. The Web is all of the computers and other applications that use the Internet, that sit on top of it so to speak, to move information around. The Web is analogous to phones and answering machines, and the Internet is like the international phone system. Applications that make up the Web, such as personal computers and email software, can continue to be improved as long as there are some standard protocols that stay in place, so that application developers can count on them being there when they design even newer Internet technologies (Berners-Lee 83). This allows for stability in development, yet means that people who design the technologies that make up the Internet and the technologies that make up the Web can produce better technologies.

Berners-Lee's call for universality of Web technology can be understood when we consider how links work. If we put a Web site online, he argues, we should be able to link to any other Web site. We also need to have technologies that people anywhere can employ regardless of the language they speak or the computer they have access to, and they need to be able to put whatever they want online. There should be no central authority that determines what goes online or who can link to whom. No one should have to ask for permission from anyone or any organization to produce a Web site and place it online.

According to Berners-Lee, for the Internet to continue thriving, we need to continue to have these "open standards" as we have had in the past (83). Because we employ "royalty-free" languages such as HTML that no one owns, anyone can use them to make a Web site without paying anyone else for the right to do so. Berners-Lee does not mean that companies cannot charge people for the services they provide in online environments. They can. That we have these open standards allows companies to be confident that they can design applications, put them online, and they will work for everyone, which allows for a greater market. Additionally, other groups of people such as scientists can produce databases of information they want to share and be assured that others will be able to gain access to it, as can governments who want to put data online so people can review it, thus fostering openness and transparency. Most notably, open technological standards allow creative people to use them in ways that had not been imagined before, in what Berners-Lee calls "serendipitous creation" (84). This creative use of technology allows us to learn from each other.

When we do not have these open standards we create "closed worlds." Berners-Lee cites Apple's iTunes as an example. Apple uses its own, "proprietary" address for the site; instead of "http" Apple uses "itunes" and this means that one needs to use the Apple's iTunes program, which is proprietary, or owned by Apple, and not open source. If you want to link to an iTunes address or to send an iTunes link to someone else, you cannot: "You are no longer on the Web. The iTunes world is centralized and walled-off. You are trapped in a single store, rather than being on the open marketplace" (Berners-Lee 83). While iTunes might be a convenient place to find and download music, it is typical of some of the "walled gardens" online today that "can never compete in diversity, richness and innovation with the mad, throbbing Web market outside their gates" (83).

Perhaps even more ominous is the practice of undermining "net neutrality." Net neutrality means that the Internet is set up for anyone to use without any "interference" (Berners-Lee 84). In the short history of the Internet and the Web, this has not always been the case. The Internet Service Provider (ISP) Comcast has demonstrated the habit of keeping others at a disadvantage by not respecting the principle of net neutrality. BitTorrent was a company with peer-to-peer network technology that allowed one user to share video and music files with another; patrons who subscribed to Comcast ISP for their Internet connection found that when they used BitTorrent the rate of exchange of information slowed or was shut down completely, thus putting BitTorrent at a disadvantage. If an ISP can do this, it can decide who can get to information and who cannot. It has the potential to allow a private company to favor companies it has business relationships with over companies it does not (Berners-Lee 84). When this happens, we do not have a level playing field in the business community and companies who have new ideas and products will have a harder time getting them to market.

The most egregious abuse of the Internet occurs when companies and governments use it for spying. Phorm is a company that developed a technology which enabled ISPs to examine the information in the packets circulating on the Internet. They could spy on people, as they could read their emails and also see all of the URLs they visited. Knowing what people look at online does not usually reveal much about them, but if it became evident that, for example, a particular person was concerned about some aspect of her health, life insurance companies could discriminate against that person (Berners-Lee and Fischetti 85).

When a government uses the Internet to read the email of people who are willing to challenge it, this undermines the ability of people in all countries to remain free. The Chinese government hacked into the email of dissidents in 2008 after Google refused to agree to its demand that some Web sites be censored from the Chinese version of Google (Berners-Lee 85).

Western governments have also been heavy handed when it comes to Internet access, by not allowing for due process of law to play out before terminating Internet privileges. The governments of France, the United Kingdom, and the United States have either disallowed Internet access or made it harder to maintain when some of their citizens have been accused of working around copyrights. For example, in France, if a person was alleged to have downloaded music or video illegally he could be denied access to the Internet for one year. Given that some people's livelihoods now depend on Internet access, this is significant. Moreover, Berners-Lee holds that the Web is so crucial to our lives that these government actions are "a form of deprivation of liberty" (85).

Exercises

1.   In Chapter 1 we discussed how different technologies through history have changed the way we communicate, think, and understand the universe and each other. In this section, the basic technologies that allowed the WWW to come into being are described, but also some of the human factors involved are discussed. What does this section say about the ethics of Tim Berners-Lee? How did his original vision of the WWW allow for its growth and adoption by so many people and institutions? Beyond the technology, what cultural elements were in place for the WWW to come into being?

2.   When you are online, how often are you in what Berners-Lee calls a "closed world"? Is it something you take for granted? Is it really the impediment to openness that Berners-Lee suggests?

HTML, XHTML, and XML

As previously described, HTML has been around for some time, and has been through a number of revisions from the original version that Tim Berners-Lee designed. Currently, the most widely used version is HTML 4.01, and it is an improvement over previous versions of HTML in that it can be used with style sheets that allow designers to better determine the layout and aesthetics of Web sites.

XHTML (extensible hypertext markup language) is based on HTML 4.01, but it also allows for the embedding of XML (extensible markup language) data descriptions in HTML code. While we will not be going into XML in much detail in this book, the conventions of XHTML will be the ones we use, as they are somewhat stricter and allow us to add metadata to our Web sites at a later date.

Briefly, XML describes the nature of data, or provides "metadata," but HTML formats it. "Metadata" is data about data. XML can tell us that the words "Moby Dick" constitute the title of a novel and that the words "Herman Melville" are the name of the author of this novel. "Moby Dick" is the data, and that it is a title of a novel is the metadata. "Herman Melville" is the data, and that he is the novel's author is the metadata.

HTML does not tell us anything about the nature of the text, but is used to describe how the text will look, where it should be placed on the screen, and how it is linked to other hypertext files. For example, it allows us to place the words "Moby Dick" and "Herman Melville" on a Web site and decide such things as the font, color, and size of the text and where it should be placed in a table. XHTML allows us to include XML data on our Web site in a database structure that conforms to a set of specified standards.

Unlike HTML, XML alone does not have anything to do with formatting. Instead, we could use XML to characterize a novel like Moby Dick, or indeed any other novel. If we had a Web site that includes a list of thousands of books, we might require that each of the entries not only describes the title and author's name, but the ISBN number, the date of the first printing, the Library of Congress number, the publisher, and the city where the publishing house is seated. We could go farther. We could say that we also need to indicate when the author was born, where he or she was born, what genre the novel is considered to fit into, and what languages, if any, the work has been translated into. If we use strict XML code, we would always be required to put this specific information in place as XML demands that we follow a template. We could do this for many things: disease classification, student records, or meteorological data. XML allows us to characterize, store, modularize, and exchange data with precision, and when we use XHTML we open up the possibility of including XML-coded data on our Web sites.

XHTML basics

For any XHTML document file, you need a basic template that includes the following features:

- Document type definition
- HTML
- Head
- Title
- Body

Document type definition

A document type definition or DTD tells the browser that all the code that follows is XHTML. Machines need to be told just what is coming up and how to read it. Additionally, it specifies what version of HTML you are using so validating software can test the syntax of your code.

There are three basic kinds of DTDs: strict, transitional, and frames. The strict version does not allow for deprecated tags, which are tags that the W3C has determined will be phased out, or tags that have already been or are slated to be phased out and have been replaced. The frames version allows for the use of framesets that are being phased out by the W3C.

The transitional version allows for some deprecated tags and frames. Below is the transitional version of the DTD:

> <!DOCTYPE html PUBLIC"-//W3C//DTD XHTML 1.0 Transitional// EN" "http://www.w3.org/TR/xhtml1/DTD/xhtml1-transitional.dtd">

While it might not be the strict DTD, it will work for our purposes as some tags we will be using might be phased out in the future but are still part of many existing Web sites. You only need the opening tag above for the DTD.

HTML

The HTML tags delimit the HTML document. They are telling the browser to recognize HTML at the opening tag, <html>, and the closing tag, </html>. This means that the browser will not display the HTML tags, but will do what the tags tell it to do to the text. For example, the "<i>and</i>" that we see in our editor will show up as the italic text *and* on the screen.

So that we are going to be producing XHTML files, we will need to enclose a namespace with the first HTML tag. A namespace identifies the particular elements and attributes that originate from a particular language, in this case, XHTML. For our purposes, we should use the following XHTML namespace:

> <html xmlns="http://www.w3.org/1999/xhtml">

Head

The head tags encapsulate other tags. These may include metatags, style tags, and comments, which we will cover later. Nothing that exists within the head tags shows up on the main browser screen. The title tags also show up between the head tags.

Character encoding for XHTML documents should be indicated with a metatag, such as the one below, which is placed between the <head> tags. The following is the particular metatag that indicates that Unicode 8 should be used:

> <meta http-equiv="Content-Type"content="text/html; charset=utf-8"/>

According to the W3C professionals, Unicode 8 is versatile as it detects the way characters (such as the letters that make up our alphabet) should be read and displayed across many different kinds of browsers and versions of these browsers.

Title

The title tags perform an obvious function; they provide the title that will show up in the title bar at the very top of the browser.

Body

All of the HTML coding for any graphics or text that we can actually see on a browser is placed between the body tags.

Review how these tags are arranged below:

```
<!Doctype>
<html>
<head>Put your metatags, style sheets, comments, and title tags
here. They will not show up on the browser screen.
<title>Title</title>
</head>
<body>Put all of your text and images here. This is what people
will see.</body>
</html>
```

Note that, other than the Doctype tag, all the other elements are encapsulated between the <html> tags.

HTML template

It is important to understand all of the concepts above because they represent the essential pieces of any HTML file. It is a good practice to always have a basic HTML template set aside as a file that you can pull up and begin putting the coding you want into; this saves you from having to type in all of the elements described above each time you produce a new file. You might use the following code for this:

```
<!DOCTYPE html PUBLIC "-//W3C//DTD XHTML 1.0 Transitional//
EN"
"http://www.w3.org/TR/xhtml1/DTD/xhtml1-transitional.dtd">
<html xmlns="http://www.w3.org/1999/xhtml">
<head>
<title>Title</title>
<meta http-equiv="Content-Type" content="text/
html;charset=utf-8" />
</head>
<body>
</body>
</html>
```

Having a file like this saved as something like "htmltemplate.html" on your computer will save you time. Make sure that when you add new information to it you save this file with: (1) a different title, such as "Emily Dickinson" and (2) a different name associated with an HTML extension, such as "emilydickinson.html."

XHTML formatting for style and layout

You have already seen the basic opening and closing tag form above, but now we need to go into more specifics about the way this affects the formatting of the information you put in your body section—the section people will see on their browsers. When you write out your HTML code, you will be putting it in tag form, which means you will be putting your elements between two angle brackets. For example, the element for italics is "i" and the tag is <i>:

> Here is her poem "Each That We Lose Takes Part of Us" from <i>Poems by Emily Dickinson: Third Series</i>.

Most HTML elements have opening and closing tags, and you can see in this example that the closing tag </i> has a forward slash before the "i." When you have an opening tag and a closing tag, you are indicating that everything that comes after the opening tag and before the closing tag is affected by the tags. In this case, the title of the book will be in italics.

Some elements only have one tag that serves as both the opening and closing tags, such as <br> for a break, or the XHTML version of break, which is <br />. Note the forward slash is after the "br" in this second version; this is accepted by the XHTML validation software that is described in a later section. This would be a tag you put at the end of a line of poetry to tell a browser to go to the next line. This can be contrasted with the opening and closing tags used below for: the paragraph element, "p," which indicates where a paragraph is to start and stop; the <b> and </b> tags for making text bold; and the "blockquote" tags that separate and indent a body of text to set it off from what comes before and after it. Again, only one tag is needed for a break, while both opening and closing tags are needed for a paragraph or blockquote:

```
<b>Emily Dickinson</b>

<p>Emily Dickinson (1830–1886) was known for her idiosyncratic
use of punctuation, line length, and meter or rhythm in her
poetry. Because of her style, only a small portion of all of the
poetry she wrote was published in her lifetime.</p>

<p>Here is her poem "Each That We Lose Takes Part of Us" from
<i>Poems by Emily Dickinson: Third Series</i> (68).</p>

<blockquote>Each that we lose takes part of us;<br />
A crescent still abides, <br />
Which like the moon, some turbid night, <br />
Is summoned by the tides. </blockquote>
```

HTML is not a case sensitive language, but XHTML is, so it is necessary to keep your elements and attributes in lower case letters. Also, as you are going to be poring over XHTML code to find where you could improve it, make changes, or correct mistakes, it is much easier on the eyes if you keep your XHTML tags in the same case. We will use lower case in this book.

Text editors

When writing any form of HTML, you will need a text editor. A text editor will allow you to save a file in plaintext, which means that no extraneous code will be added to the code you write and therefore the machine that eventually reads your work will read and display it correctly. You cannot use word processing software such as Word to write out your code, as it will add such extra coding to your HTML documents and render them unreadable on a browser; you need to use a text editor.

There are many text editors available that are free, and many are included with computer operating systems. Perhaps the most widely used is Microsoft's Notepad, which comes with Windows operating systems. If you own a Mac, you can use TextEdit or SimpleText, as they are included with different versions of Mac operating systems. There are other text editors that have syntax highlighting features that make it easier for you to see your code as you write and alter it. Some are free for downloading.

If we were to take all the code we have been discussing above and combine it for our first Web site, we have the following in the editor Notepad:

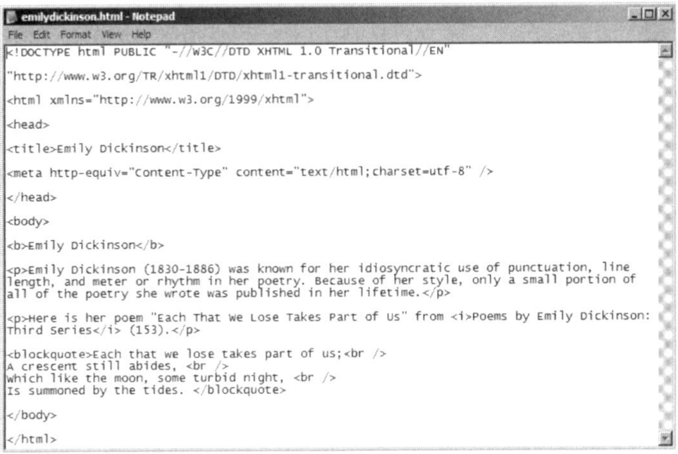

Figure 2.1 HTML file in a text editor.

Note how all of the basic code described above is included here: the DTD, the XHTML namespace, and the character encoding information are all in place and will be reused in the other files presented in this book.

Saving our files

Now we have to save our file with an HTML extension (.html). To do this, perform the following sequence of actions:

1. Pull up Notepad® or any other editor of your choice on your computer and write in the text shown in the screenshot above.
2. Go to "File" in the upper left-hand corner of the Notepad screen, then hit "Save As."
3. Make sure that you get rid of any information in the "File name" box that is already there, such as "*.txt." This box should be empty.
4. Type "emilydickinson.html" in the file name box. Make sure that in the "Save as type" box it says "Text File."
5. Hit the "Save" button in the lower right of the Notepad® screen. Make sure you save the file in a place where you can find it. For starters, you can just save it on your desktop.

Below is a screenshot of this procedure using Notepad:

Figure 2.2 Saving the HTML file in a text editor.

After you have saved this file in an editor, perform the following two steps:

1. Pull up a browser such as Internet Explorer or Mozilla Firefox.
2. Go to "File" in the upper left hand corner, and depending on the browser you have chosen, choose "Open" or "Open File." Find your file, emilydickinson.html, and select it. If you pull this up in a browser such as Internet Explorer, you get the following:

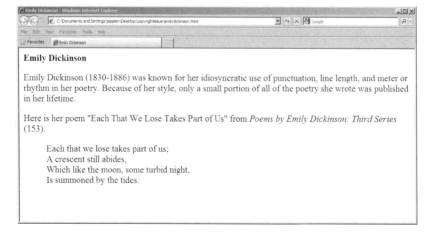

Figure 2.3 HTML file shown in a browser.

As you look at this file, note how the <p>, <i>, <b>, <blockquote>, and <br /> elements changed the plain-looking text seen in the text editor to the text seen in the browser. You can see how the formatting created separate paragraphs and where the italics are used to indicate a book title. The poem is separated from the one-line paragraph above it in a block, and the lines of the poem break where they are supposed to.

Additionally, at the very top of the browser, the title "Emily Dickinson" shows up before "Windows Internet Explorer." This title is in place because we put <title>Emily Dickinson</title> between the <head></head> tags in the HTML file.

Additional HTML code tags

In addition to these five basic HTML tags—<p>, <i>, <b>, <br />, and <blockquote>—it is important to know some of the following basic tags. The <i></i> tags that are used for italics can be substituted with <em></em> or "emphasis" tags. Additionally, <strong></strong> tags can be used in place of the <b></b> tags that indicate bold.

This is not an exhaustive list and we will describe other HTML tags in the pages that follow.

Unordered list

An "unordered list," or a bulleted list, is used to convey key points about a subject. We could illustrate a series of closely related but separate ideas in a paragraph, or we could separate them into their own distinct units so they are easier to read and understand.

As in the example below, note that we always start and end unordered lists with the opening and closing <ul> or unordered list tags. At the beginning and end of each item in the list, we use the <li> and </li> tags respectively.

```
<p>In addition to HTML, Tim Berners-Lee needed to invent the following technologies: </p>
<ul>
<p><li>URIs—"Uniform resource identifiers" or URIs, which we now call "uniform resource locators" or URLs, are the addresses for Web sites that allow us to find a site on a server and review it without having to write complex computer programs to find and access the information on other computers. </li></p>
```

```
<p><li>Browser—A browser decodes a Web site's HTML and
allows us to read it. </li></p> <p><li>
Web server—This is the software that allows us to store informa-
tion on a computer and allows others to access it. The first server
was on Berners-Lee's personal computer and was registered as
info.cern.ch. </li></p>
<p><li>HTTP—Protocols are basic rules that allow computers to
exchange information between each other. "Hypertext Trans-
mission Protocol" is a simple protocol language that allows Web
browsers to find and identify the URIs of HTML-based Web sites
on servers and translate the HTML.</li></p>
</ul>
```

Here is how it would look in a browser:

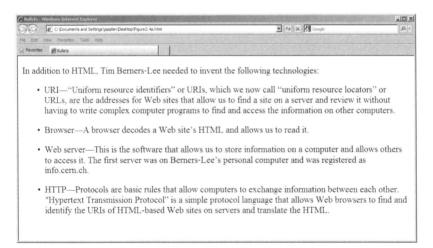

Figure 2.4 Unordered list in a browser.

Note that each individual bullet is separated from the others and they are indented. They are separated because we used the <p> tags around each line, so we get a space.

In place of the default "black circle" bullets, we can also use other styles, such as squares and circles:

```
<ul>
<li type=square>This is a square,<br />
<li type=circle>and this is a circle.
</ul>
```

Here is how it would look in Internet Explorer:

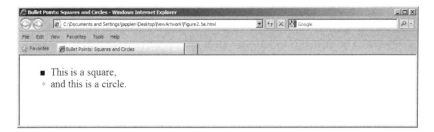

Figure 2.5 Alternative unordered list in a browser.

Ordered list

An ordered list is an enumerated list. It describes the order of the steps we need to take in performing a task. Unlike a bulleted list, an ordered list usually describes a sequence of actions or a process. Note that ordered lists read better if the first word in each sentence is an active verb or part of an active verb phrase; "make sure," "go to," and "type" are used in the code below.

To begin the list, we start with the <ol> or ordered list tag and close with the </ol> tag. As in the unordered list above, we open and close each item in the list with the <li> and </li> tags.

```
<h2>Saving Our Files</h2>
<p>Now we have to save our file with an HTML extension (.html).
To do this, perform the following sequence of actions:</p>
<ol>
<li>1.Pull up Notepad® or any other editor of your choice on your
computer and write in the appropriate text.</li>
<li>2.Go to "File" in the upper left-hand corner of the Notepad®
screen, then hit "Save As." </li>
<li>3.Make sure that you get rid of any information in the "File
name" box that is already there, such as "*.txt." This box should
be empty.</li>
<li>4.Type "emilydickinson.html" in the file name box. Make sure
that in the "Save as type" box it says "Text Documents." </li>
<li>5.Hit the "Save" button in the lower right of the Notepad®
screen. Make sure you save the file in a place where you can find
it. For starters, you can just save it on your desktop.</li>
</ol>
```

Below is the way our ordered list would look in Internet Explorer:

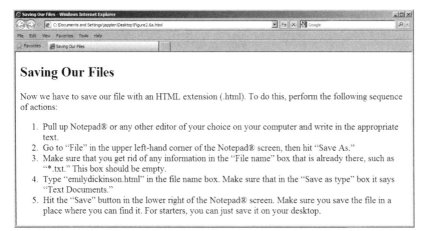

Figure 2.6 Ordered list in a browser.

In this unordered list, we did not use the <p> tags around each separate list item, so we did not get the spacing between each item on the list. Also note that we use the <h2> tag to get a larger heading than the body of the text. We will discuss headings next.

Headings

Basic HTML makes it very easy to use headings of varying sizes, and it is easy to manipulate these sizes. In the sample above, we have h2 tags surrounding the heading "Saving Our Files," and it shows up in the same font as the text that follows it.

There are six heading tags: h1, h2, h3, h4, h5, and h6. H1 is the largest heading, and h6 is the smallest, which might seem counterintuitive. What these heading tags do is take whatever font we are using, change the size, and make it bold, thus making it easier for us if we need to employ different levels of headings and subheadings.

Below is the code for the six headings and one regular sentence:

```
<h1>Heading One</h1>
<p>Now is the time for all students to come to the aid of their
universities.</p>
<h2>Heading Two</h2>
<h3>Heading Three</h3>
```

```
<h4>Heading Four</h4>
<h5>Heading Five</h5>
<h6>Heading Six</h6>
```

This is how these different headings would look in a browser:

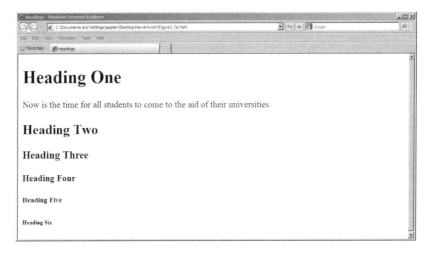

Figure 2.7 Headings in a browser.

The line "Now is the time for all students to come to the aid of their universities" is the regular default text, and the different-sized headings are in bold but take the same font style as this text.

Elements, attributes, and values

"Elements" identify the specific layout, navigation, or style convention that you want to be in place when you code your Web sites. If you think of yourself as an editor who is marking up the text of a manuscript, when you are using the element for italics, you are saying, "Put this title in italics."

"Attributes" tell us something specific about the element we are putting in place. For example, if we want to put an image in our Web site, we would use "img" for the element but we would have to furnish some additional information or attributes, such as the source, or "src," the height and width of the image, and the border around it:

```
<img src="tbl2001.jpg" height="197" width="300" border="0"/>
```

These attributes are telling us more than "there is an image to be placed here." They tell us that there is a source name, a height, a width, and a border.

"Values" are even more specific, as they tell us that the name of the image is "tbl2001.jpg," and that the height, width, and border are "300," "197," and "0" pixels respectively.

Validating your HTML code

Even if your Web site seems to look fine on one browser, it is good practice to validate your Web site, and the best way to do this is to use the online validating software of the W3C. Below are the reasons for validating your HTML code:

- Validation allows you to debug your code. While sometimes browsers are forgiving and correct or overlook imperfections in your HTML, sometimes they do not. With different browsers and several versions of each browser on personal computers throughout the world, having HTML-validated code is the best way to ensure that people will be able to view your Web sites the way you want them to. Just because something looks great on the browser you are using does not mean it will look great on another browser or platform, or on either a PC or a Mac.
- Validation offers the best insurance that newer browsers in years to come will also read your code, as they are generally designed to meet W3C standards.
- Validation ensures a standard in coding and will allow more people to work on and maintain a Web site because they all know the standard.
- Validation teaches good HTML practice from the beginning and makes your Web sites look professional. When people select "Source" or "Page Source" and look at the code of your Web site, it will reveal a competent technician.

You can go to the W3C validation site at http://validator.w3.org/#validate_by_upload+with_options and validate a Web site that is already online or your HTML files. Also, you can just type in some markup code you are concerned about and have it checked.

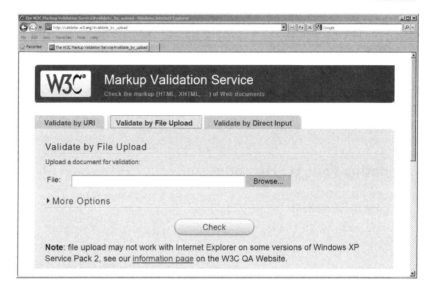

Figure 2.8 The W3C HTML validator.

Before you upload your Web site to a server, you can check each file using the "File Upload" tab on the validation site to see if your coding is meeting contemporary standards. You would select "Browse" and go through your hard drive or any other drive that contains your files, select the HTML file you want to test, and select "Check."

Additionally, you can validate your CSS code, which we will be covering later, using the W3C CSS validator: http://jigsaw.w3.org/css-validator/validator.html.en#validate_by_upload.

Deprecated HTML tags and HTML5

At the time of writing, HTML5 has been coming into play and Web site coders will be adopting it. Deprecated HTML tags are those tags that were used in previous versions of HTML, but which are not necessarily "supported" by the newer browsers. For example, the tag for centering text in HTML 4.01, <center>, has been deprecated and is being replaced in HTML5 with <div style="text-align:center">. That is:

    Centered Heading

will be replaced with:

    Centered Heading

A more common trend is to use external CSS to do things like center text. We will cover this in Chapter 3.

To understand the process that produces these changes in tags, we need to understand the culture of the Web and how the recommendations of the W3C come into play. The W3C, a non-profit organization, brings Web site architects, HTML/CSS enthusiasts, and representatives from software companies together to see what might be problematic about some of the tags that are used on the Web, then the W3C suggests changes. If all or most of the organizations who produce browsers such as Mozilla Firefox, Safari, Internet Explorer, or Chrome build the newest versions of their browsers to recognize, decode, and display texts in the fashion that the new tags were designed for, these new tags will work or be "supported" and will probably become a new standard.

What is also important is that people who build Web sites use these new tags in their code. Thus we have a "socially constructed phenomenon" (which is discussed in more detail in Chapter 4); when the browsers decode tags and people see this, they continue to use these tags and teach others to use them, and thus they become the HTML "vocabulary" with which we can all converse. The W3C can only recommend new tags, not make people use them, but if browsers decode only some tags and not others, Web sites will not be rendered well and people will stop using these browsers in favor of others.

Deprecated tags are usually recognized for years after they have been formally taken off the W3C list of acceptable tags, as many older Web sites still use them. If browsers do not read some of the older Web sites, people will most likely use other browsers. It is important to keep in mind that when a new version of HTML comes out, really very little changes; the vast majority of HTML tags that were in use well over two decades ago are still used and supported.

First HTML exercise

Pull up an editor such as Notepad or Textedit, and type in the code shown below that is marked in gray (note there are two parts, at the beginning and end of the code). Leave a space between the <body bgcolor="E0E0E0"> and </body>. Save the file as "firsthtmlexercise. html." Make sure you follow the directions for saving; you will be using an HTML extension (.html).

If you type the code in correctly, save it, then pull it up in a browser, you will get a blank screen with a light gray background. In the browser's title bar at the very top of your screen you should see "First HTML Exercise."

```
<!DOCTYPE html PUBLIC "-//W3C//DTD XHTML 1.0 Transitional//
EN"
"http://www.w3.org/TR/xhtml1/DTD/xhtml1-transitional.dtd">
<html xmlns="http://www.w3.org/1999/xhtml">

<head>

<title>First Hypertext Exercise</title>
<meta http-equiv="Content-Type" content="text/html;
charset=iso-8859-1"/>

<style type="text/css">
p.dropcap:first-letter {font: 200% times; float: left; margin-right:
5px; margin-bottom: 1px;}
.image_left {float: left; margin-top: 10px; margin-right: 8px;}
p.indent {font-family: Calibri, Optima, Arial, sans-serif; text-
indent: 25px; margin: 5px 30px 0px 0px; font-size: 1.0em;}
p:first-letter {font-size: 180%; color:#9933cc;}
table {margin: auto;}
</style>

</head>

<body bgcolor="E0E0E0">

<div style="text-align:center">
<table border="0px" cellspacing="0px" cellpadding="0px"
width="60%" bgcolor="#FFFF10" margin: auto; >
<tr><td><h1>Writing for the Web</h1></td></tr>
</table>
</div>

<table border="0px" cellspacing="0px" cellpadding="0px">
<tr> <td> </td> </tr>

<tr><td bgcolor="#3366FF" style="padding-left:10px"><font
color="#FFFFFF">The advent of <b>hypertext</b> is having a
considerable impact on our society; much if not most of <i>our
future learning, communication, and information gathering
activities</i> will utilize the World Wide Web.</font></td>
<td ><img src="purpleflower.jpg" height="190px" width="300px"
alt="Purple Flower"/></td>
```

<tdbgcolor="#3366FF"><pclass="dropcap"><fontcolor="#FFFFFF">
Writing for the Web is a writing and Web page construction
course, which means you will be learning how to use the library
and the Internet to gather, distill, and present information in two
media: a research paper and a web page. There will also be a
number of shorter writing assignments and journal summaries in
the course, and we will be spending a <em>significant amount of
time</em> studying theories about information so we can
develop a more <strong>critical sensibility</strong> regarding
hypertext.</font></p></td>
</tr>

<tr> <td> </td></tr>
</table>

<p>The topic for your paper will be one of the many issues
associated with hypertext, issues such as the following:</p>
<ul>
<li type="square">First Amendment rights and copyright law;</
li>
<li type="disk">international politics;</li>
<li type="circle">and the use of the World Wide Web in business,
law enforcement, and education. </li>
</ul>

<p class="indent" style="color:sienna">After you have finished
your written project, you will reformat your ideas in HTML (hypertext markup language) and CSS (cascading style sheets) and
produce a Web site so you can better understand the advantages
and disadvantages of both media.</p>

<blockquote>This does not mean that every word in your paper
will find its way onto your Web page, nor am I suggesting that you
cannot add other elements to your Web page such as graphics
that were not in your paper.</blockquote>

<img src="menontrain.jpg" height="132px" width="200px"
alt="Men on Train" class="image_left" />

<p><font size="15px" color="#FF0000">In the last week of the
course you will present your Web page to all of us.</font> Your
final Web page will be ported to the

```
<a href="http://www.english.ucf.edu">Department of English
</a> Web site.</p>

</body>

</html>
```

After you have done this, begin typing in the code between the parts marked in gray, saving it, and then pulling it up in a browser such as Mozilla Firefox, Safari, or Internet Explorer. Each time you save the file and then pull it up, more text will come into view.

If you have typed everything correctly, you should end up with a screen like this:

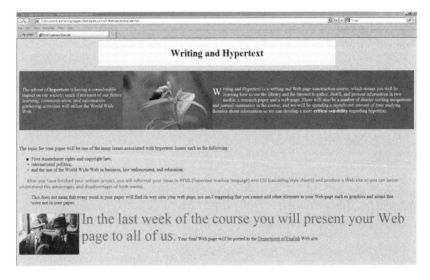

Figure 2.9 The first HTML exercise in a browser.

To get the two images—the flower and the two men on a train—to render, you will need to pull them off the Web site for this book. Locate the two files "smallpurpleflower.jpg" and "menontrain.jpg" on the site and copy them to the folder that contains your HTML file. They all need to be in the same folder or the two images will not show up. You can place all three files—the one HTML file and the two .jpg files—on your desktop. This counts as a folder. You can also make a separate folder for them.

After you have finished this exercise, look at the text on the screen and then look at the HTML code around it in the editor. How does the code affect the text and placement of the images? You can play around with the code and see how changes affect the look on the screen. Some examples are provided in the exercises that follow.

Exercises

1. Go online and find some "Web-safe" color codes. Find a different "hex RGB" color code combination than #3366ff and replace it. For example, try #7fffd4, the hexadecimal code for "aquamarine." The W3C has a page dedicated to these color codes: http://www.w3.org/TR/css3-color. There are many other Web sites that also have this information.
2. Change the font size from 15px to another value. How does this affect the look of the text on the screen?
3. We have some document level CSS coding between the <head> and </head> tags. Change the value of the p.indent selector from 25px to 35px and see what you get. We will be learning more about CSS in the chapters to come, but it is good to get a feel for it now.
4. Go to the CSS validator referenced in the last section and upload your file. If you have typed all of this code in correctly, you should get a message that states: "This document was successfully checked as XHTML 1.0 Transitional!"

Links

If there is one overall character that we can ascribe to hypertext technology that makes it unique, it is the way documents are linked together. If you think about it, when you are looking at any one page of a hypertext document, it could also be replicated in print form. However, that we can so readily move from one document to another and move across large bodies of hypertext documents with such ease is what really gives hypertext its power.

There are two basic kinds of links: absolute and relative. When we make a relative link, we are linking from one HTML file on our Web site to another file on our Web site. When we make an absolute link, we are linking outside our own HTML files to another set of HTML files that are on another Web site. For example, a link from the opening of your own Web site to your resume which you also have on your own Web site is a relative link. A link from your personal Web site to the Web site of an organization you belong to, such as the Society for Technical Communication or the Sigma Tau Delta society, is an absolute link.

You have probably noticed that companies that sell us things in online environments rarely have absolute or outside links, as they want to keep us "in the store" or on their Web site so we do not get distracted and go to other Web sites and perhaps forget to buy something. On Web sites that are more about providing information than selling something, absolute links can be used to augment or support the position or ideas of those who are producing the Web site, so we are more likely to see absolute links there. However, when you provide a link to another Web site, remember that people often go to this site, and perhaps beyond to other Web sites and away from what you have produced.

The HTML code for a relative link is as follows:

<a href="nameofyourfile">name of your file</a>

The "a" stands for "anchor," and the "href" stands for "hypertext reference." Additionally, note that there should be no spaces between words in your file name. For example, "name of your file" would not work.

For example, if you were linking from your home page to your resume, which is a separate HTML file in your Web site, you might use the following:

> If you would like to know more about my education and professional background, you can review my <a href="resume.html">resume</a>.

This assumes that your resume has been coded in HTML, that you have named it "resume.html," and that it is in the same folder as the page from which you are making the link. It is best that you have your link embedded in text as opposed to saying something like "Click here to see my resume" with the "Click here" underlined and in blue and serving as a link. This is an outdated practice. When the Web was a relatively new phenomenon, people really did not know that text that was underlined and in the default blue used for links actually suggested another linked body of information out in the WWW that they could go to.

If you have a link to an outside source you would write <a href="URL">URL</a>. For example, for a link to the Society for Technical Communication, you might embed the absolute link in the following sentence:

> You will see that I have been a member of the <a href="www.stc.org"> Society for Technical Communication</a> for three years.

If you were to put these sentences with the two different kinds of links together, you would have the following:

Figure 2.10 Absolute link in a browser.

Note how the links stand out for the viewer to use if he or she wants to, but are embedded in sentences.

In the example of an absolute link above, the URL used is the Society for Technical Communication home page, and this is where you would land if you selected it. However, if you were making a link to another page that is part of the STC Web site, such as a membership page that resides "below" the home page, you would be making a "deep" link. Deep links can get people to a precise point quickly, and as sites become more complex this is important.

Here is a simple exercise in which we will make a simple relative link from one page to another. Adding to the basic HTML template, put the following text and HTML code between the <body> tags:

> This is my first file. Do you want to see <a href="secondfile.html">my second file</a>?

Save this file as "firstfile.html." Now make a second file that uses this text and code:

> This is my second file. Do you want to see <a href="firstfile.html">my first file</a>?

Save this file as "secondfile.html."

When you pull these files up in your browser, they should look like this:

Figure 2.11 A first file with a relative link to a second file.

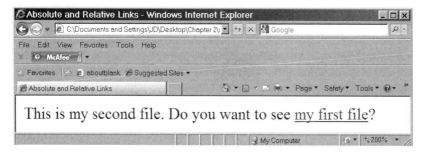

Figure 2.12 The second file with a relative link to the first file.

If you click on the link in the respective HTML files, you should be able to go back and forth between them.

Note: It is important that you keep both of these files in the same folder. The folder can be your desktop, or it can be a folder that you make specifically for these two files to separate them from all other files.

Files, folders, and pathways

It is important to save our HTML files in a separate folder because this allows us to be organized. Also, our separate files can be linked with greater ease if they are in the same folder.

Today's operating systems on both PCs and Macs make this easy. For example, below is a screenshot of a folder of HTML files for a personal Web site: a home page, a brief biography, a set of writing samples, a resume, and a set of personal links to other Web sites. In addition, there is a graphic entitled "purpleflower.jpg":

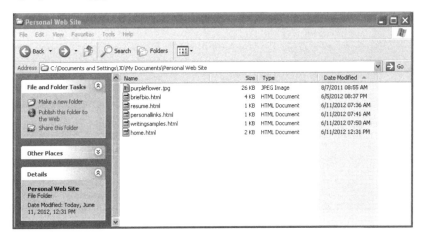

Figure 2.13 HTML and image files in one folder.

Because these files are in one folder, they can be linked with the simplest HTML syntax. For example, if we want the "home.html" file to link to the "resume.html" file, we would write this code in the "home.html" file:

```
<a href="resume.html">Resume</a>
```

This line of HTML code would produce a simple link entitled "Resume" on your home page. If the "resume.html" and the "home.html" files were in separate folders, or if one was on your desktop and one was in another folder, they would not link with this code.

If we begin adding HTML and graphics files to our Web sites, it is often a good idea to organize them by creating folders within the primary folder, in this case the "Personal Web Site" folder. For example, we could put two of the files, "resume.html" and "writingsamples.html," into a separate folder entitled "professional." Additionally, it is a common practice to separate HTML files from graphics files—with extensions such as .jpg and .gif—by placing them in their own folder, a folder we would call "images." If we did this with the set of files above, we would have a primary folder that looked like this:

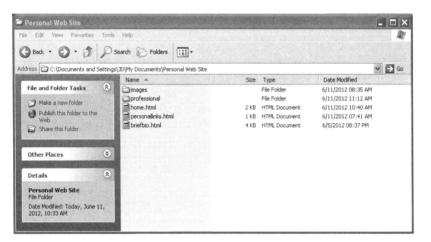

Figure 2.14 Separate HTML and image file folders.

Because "resume.html" is in a separate folder than the "home.html" file, linking these two files needs an extra step. To describe the pathway from the "home. html" to the "resume.html" file requires that we identify the separate folder that "resume.html" is in, "professional," and thus we use this code on the home page:

    <a href="**professional**/resume.html">Resume</a>

The folder "professional" is referenced first, followed by a forward slash (/), followed again by "resume.html."

In the later section in this book on how to place an image in a file so it shows up on the screen, the coding for the "purpleflower.jpg" file is given as:

    <img src="purpleflower.jpg" height="190" width="300" alt="Purple Flower"/>

We would use this code if the "purpleflower.jpg" file were in the same folder as the other files, the HTML files that we want it to show up on. However, if we had it saved in the "images" folder, we would use this code:

    <img src="**images**/purpleflower.jpg" height="190" width="300" alt="Purple Flower"/>

Note that the "images" folder is referenced first, followed by a forward slash (/), then "purpleflower.jpg." The forward slash indicates that, to find and link to this specific image, the browser has to move one folder up on the pathway.

Mailto link

Whenever you want to leave an email address on a Web site in link form, as opposed to showing the actual email address on the page, you would write out the following:

<a href="mailto:youremailaddress@example.com">Your Name</a>

A link entitled "Your Name" would show up on the file where you placed this code. To be clear, the "youremailaddresss@example.com" is a made up email address, and the distinguishing part of this linking syntax is the "mailto" and colon.

Image maps

Image maps are images that you can drag a mouse over and see a small pointing hand icon show up in certain regions that, when selected, link you to other absolute or relative files.

Image maps can be "server-side" image maps, which reside on a server, or "client-side" image maps, which do not. Because the way that servers process image maps varies, there can be a delay in their appearing on screen. Client-side image maps avoid this.

The following are the basic tags for image maps:

map—The opening and closing tags describe the image map specifications. These include the name of the map and the map id. They lie within the body tags.

area—This tag delineates the area on the image where the mouse-sensitive icon is activated. This includes the specific regional map shape and coordinates of the mouse-sensitive area. It resides within the opening and closing map tags.

shape—The shape of the area is described by this tag. It can be a rectangle, a circle, or a polygon, indicated in the shape tag code by "rect," "circ," and "poly," respectively.

coords—This gives the x- and y-coordinates within the overall image that delimit the boundary of the shape. The values given for the coordinates vary depending on the shape, as follows:

o   **rectangle**—The coordinates given for a rectangle are x1 and y1 for the upper left corner and x2 and y2 for the lower-right corner. The formula for a rectangle is "x1, y1, x2, y2." In the Mount Rushmore example below, "22, 11, 90, 80" gives the coordinates for these two corners or points on a rectangle covering George Washington. For a rectangle, you only have to indicate these two points. How to determine these coordinates on an image is discussed below.

o   **circle**—The first two coordinates, x and y, define where the center of the circle is positioned on the image map. The z value is the length of the radius that extends outward from the center point. Thus we have "x, y, z" as the formula for a circle.

o   **polygon**—A polygon is indicated by naming the coordinates for every corner of the shape. Thus we would use "x1, y1, x2, y2, x3, y3, ..." for as many coordinates as we need.

alt—This is an area element that allows those who cannot see the image to at least have a verbal description of it. This can be the case for someone who is visually impaired or for a browser or other technology that does not render, or display, the image. It is included in the area and img src tags.

usemap—This indicates that the code provided is for a client-side image map and establishes a connection between the map and the graphic. In our example, the usemap name is the same as the map name, but it is preceded by a #. It is enclosed in the img src tag.

img src—This designates the image that should be in place and whether it is a .gif or a .jpg, or another kind of file. The area, shape, coords, alt, and usemap tags designate values that are associated with the specific image referenced by this tag. This tag lies outside the map tags but within the body tags.

Below is the source code for a simple image map using an image of Mount Rushmore:

```
<body>

<map name="rushmore" id="rushmore">

<area alt="Washington Monument National Park Service Web
site" shape="rect" coords="22, 11, 90, 80" href="http://www.
nps.gov/wamo" />

<area alt="Lincoln Memorial National Park Service Web site"
shape="rect" coords="214, 53, 279, 131" href="http://www.nps.
gov/linc" />

</map>

<img src="rushmore.jpg" alt="Mt. Rushmore"
usemap="#rushmore" width="285" height="180"/>

</body>
```

Here is how it looks in Internet Explorer®:

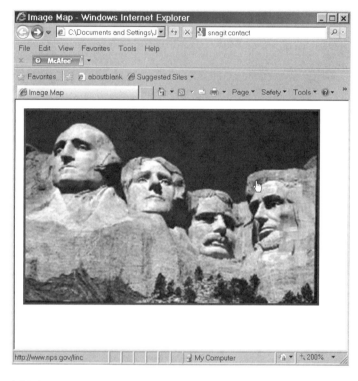

Figure 2.15 A Mount Rushmore image map in a browser.

As you can see, the pointing hand icon appears in the designated area on Lincoln. When selected, we are taken to the Lincoln Memorial Web site:

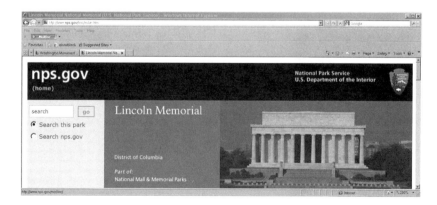

Figure 2.16 Lincoln Memorial absolute link.

In addition to the name of the file, you will also need to know the size of your graphic in the "img src" tag. Most graphics editing programs will allow you to find this. Below is the image size window from Adobe™ Photoshop™, where we can see that the width of the Mount Rushmore image is 285 pixels and the height is 180 pixels.

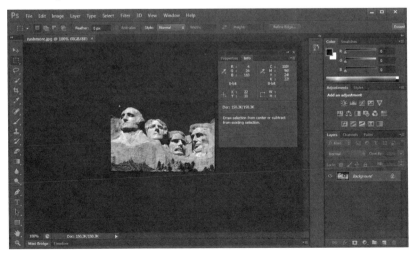

Figure 2.17 Image size dialogue box.

To find the coordinates of the region in the image map that the mouse-sensitive hand icon will activate in, you can use the "Info X" box in Adobe™ Photoshop™:

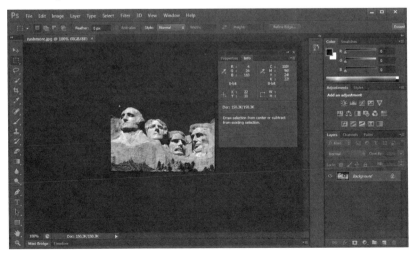

Figure 2.18 Coordinate selection dialogue box.

The white "+" cursor in the upper left corner of the region around George Washington's image indicates the position of the upper-left hand corner of a rectangle. The *x*- and *y*-coordinates of the corner, 22 and 11, can be seen in the Info X dialogue box. Make sure that you have the coordinate values set in pixels. You can do this by selecting the black "+" to the left of the "X" and the "Y" and then choosing pixels (as opposed to inches or something else):

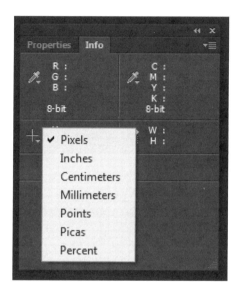

Figure 2.19 Coordinate measurement choices.

Creating a good image map requires that the areas on it that you define with coordinates allow the patron a sense of just what might show up when he makes a selection. For example, selecting the area over the representation of Abraham Lincoln on the Mount Rushmore image leads us to the Lincoln Memorial Web site. If we had a map of the United States and you made a selection over San Francisco, you would expect go to some Web site associated with that city.

An image map is another way of organizing information. It illustrates a connection between things that are organized across space or geography. Imagine a list of simple <a href> text links to information on all your favorite American cities, versus an image map of America with the same cities identified on it. In this case, an image map suggests much more than the text links. See, for example, the image map of the Mayo Clinic in the section on "Web Site Architecture" in Chapter 5.

As not all image maps are completely self-explanatory, it might be a good idea to label the areas on the map that indicate a selectable area; for example, see Brittney Adams' image map on her personal Web site, shown in Chapter 6. Using the "Horizontal Type Tool" in Adobe ™ Photoshop™ is one way to do this.

Exercises

1.   Change the shape of the George Washington or Abraham Lincoln regions to a polygon or a circle. You will have to find the coordinates for these using a graphics program.
2.   Add two more absolute links to the code, one to the Theodore Roosevelt National Park Web site (http://www.nps.gov/thro/index.htm), and one to the Thomas Jefferson Memorial Web site (http://www.nps.gov/thje/index.htm). You will have to find the coordinates for the appropriate designated areas using a graphics program. By the end of this exercise, you will have an image of Mount Rushmore with four image map connections, one for each president.

Anchor links

Anchor links allow us to link to a specific section within an HTML file. If we are linking within files that are rather short, the reason for a link is apparent when we select it; the information that is described by the blue underlined link shows up at the top of the page that the link takes us to. However, some files are rather long and one might have to scroll down to find specific information, which might not be convenient. Anchor links point us to specific parts of a file so we do not have to scroll up and down to find the information we are looking for.

The "source" anchor is the link we first select on the page, and the "destination" anchor takes us to the information that shows up when we arrive at our destination.

If we wanted to have a list of source links at the top of a page and a series of destination anchors on the same page, the formula for a source link is as follows:

```
<a href="#anchor">source link text that indicates anchor below</a>
```

The hash, #, is directing the browser to a specific place on the page, not just the top of the page, which is the usual outcome for regular absolute and relative links. For the destination anchor, we use the following syntax:

```
<a name="anchor">destination anchor</a>
```

You do not have to use "anchor" in either formula; this is just a convenient way to describe it. Use a name that best indicates to you the place that the viewers will be able to jump to.

Below is the code for a one-file version of Abraham Lincoln's Gettysburg Address, and some attendant commentary below it:

```
<h2>The Gettysburg Address</h2>

<a href="#link4">Four score and seven years ago</a> our fathers
brought forth on this continent a new nation, conceived in liberty
and dedicated to the proposition that <a href="#link2">all men
are created equal</a>. Now we are engaged in a great civil war,
testing whether that nation or any nation so conceived and so
dedicated can long endure. We are met on a <a href="#link5">great
battlefield of that war</a>. We have come to dedicate a portion
of that field as a final resting place for those who here gave their
lives that that nation might live. It is altogether fitting and proper
that we should do this. But in a larger sense, we cannot dedicate,
we cannot <a href="#link3">consecrate</a>, we cannot hallow
this ground. The brave men, living and dead, who struggled here
have consecrated it far above our poor power to add or detract.
<a href="#link7">The world will little note nor long remember
what we say here</a>, but it can never forget what they did here.
It is for us the living rather to be dedicated here to the unfinished
work which they who fought here have thus far so nobly
advanced. It is rather for us to be here dedicated to the great task
remaining before us&mdash;that from these honored dead we
take increased devotion to that cause for which they gave <a
href="#link6">the last full measure of devotion</a>&mdash;that
we here highly resolve that these dead shall not have died in
vain, that this nation under God shall have a new birth of free-
dom, and that government of the people, by the people, for the
people shall not perish from the earth.

&mdash;<a href="#link1">Abraham Lincoln</a>

<h3>Glossary and Commentary</h3>

<p><a name="link1"><b>Abraham Lincoln</b></a>: The sixteenth
president of the United States. More books have been written
about Lincoln than any other historical figure in American history.
```

</p>

<p><a name="link2"><b>"all men are created equal"</b></a>: This phrase was first written by Thomas Jefferson in the Declaration of Independence: "We hold these truths to be self-evident, that all men are created equal, that they are endowed by their Creator with certain unalienable Rights, that among these are Life, Liberty and the pursuit of Happiness." </p>

<p><a name="link3"><b>consecrate</b></a>: To declare or set apart as sacred. Lincoln is lifting this battle and its participants to a higher level here. </p>

<p><a name="link4"><b>"Four score and seven years ago"</b></a>: Eighty-seven years ago. Score means twenty, so four times one score equals eighty and then you add seven to get eighty-seven. Four score and seven gives this speech a little more gravitas, as if Lincoln was reaching farther back into history than just eighty-seven years to when the Declaration of Independence was signed. </p>

<p><a name="link5"><b>"great battlefield of that war"</b></a>: <a href="http://www.nps.gov/gett/">Gettysburg, Pennsylvania</a>, and the surrounding countryside. </p>

<p><a name="link6"><b>"the last full measure of devotion"</b></a>: They died in battle for the preservation of the Union, and "the last full measure of devotion" sounds better and perhaps more appropriate than "they died." </p>

<p><a name="link7"><b>"The world will little note nor long remember what we say here"</b></a>:<a href="#link1">Lincoln</a> is being rather modest here; every American has heard or read this speech.</p>

Note how the different anchor links, the source anchor links in Lincoln's address and the destination anchor links in the commentary, are used.

Here is how it looks in a browser:

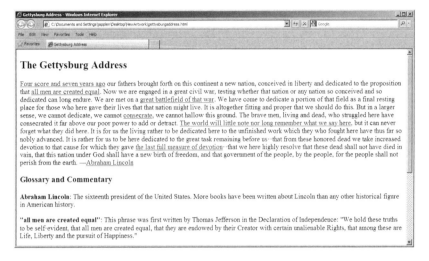

Figure 2.20 The "Gettysburg Address" and commentary in a browser.

It is easy to see the first three complete entries in the glossary and commentary section. If one was to select the link to "the last full measure of devotion," the screen would scroll downwards automatically and we would see the following:

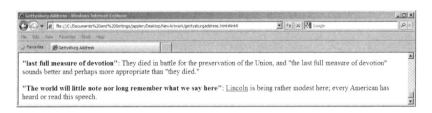

Figure 2.21 Scrolled-down glossary.

Anchor links are also used to allow viewers to jump to a specific part of another file or Web site.

For example, we can split up the single file containing both Lincoln's speech and the glossary into two separate files that are still connected with anchor links. Below is a source link that would be used in a file that contained only Lincoln's address. It both embeds some text and points to a relative link, "gettysburgglossary.html":

```
<a href="gettysburgglossary.html#link6">the last full measure of
devotion</a>
```

All we are doing here is naming the HTML file that contains the glossary and then adding the anchor name with a hash.

The formula for the destination link is as follows:

<a name="anchor">text on destination page</a>

Here is the destination link that designates text in a second file containing the glossary:

<a name="link6">"the last full measure of devotion"</a>

You do not have to use "link 6" for the name. This is just a convenient way of describing it. When using this method, you need to think of a name that will allow you to make sense of the order of your anchors.

If we did this using the material above, we would have two files. One file would be The Gettysburg Address, and the other would be the Glossary and Commentary section. The two separate files for these texts, "gettysburgad-dressanchorlinks.html" and "gettysburgglossary.html" are on the Web site for this book. These are written as separate files with relative links.

Tables

The tags we use to set up tables in the HTML world were originally con-ceptualized as a means of presenting data and replicating the conventional use of tables in science, business, government, and other fields. Tables allow us to present many data points easily, so we can see the relationships between, for example, date, temperature, and precipitation or rainfall measurements. However, we have found that tables also allow us a means of organizing information such as bodies of text and images in layout patterns we deem to be appropriate for our Web sites. Both uses of tables—for data and layout organization—will be explained below.

Using tables to display information

Here are the basic HTML elements for building tables on Web sites:

Table—The most important tag is <table>. Whenever you make a table that is separate from other information, you start with this tag.
Table row—The table row tag or <tr> is just that, a row that goes left to right.
Table division—The <td> or table cell tag can be thought of as the tag that allows us to divide up each table row into smaller units or cells. Perhaps it is easier to think of it as a "table division" tag as we use "td" in the tag. As there is no tag for table columns, what we do in HTML is to stack table

rows on top of each other, then have divisions within each table row that when stacked up look like columns.

Table header—If we want to put a header at the top of each column to label the contents, we do this with a table header or <th> tag.

In the code below we are using all of these tags to create a simple table with two columns, one row that consists of headers or labels, and one row beneath it. This includes a border width of 2 pixels, thus we have <table border="2"> for the opening table tag:

```
<table border="2">

<tr>
<th>First Column Label</th>
<th>Second Column Label</th>
</tr>

<tr>
<td>First Cell, First Row</td>
<td>Second Cell, First Row</td>
</tr>

</table>
```

Because we have specified a border width of 2 pixels, we can see the outlines of the table:

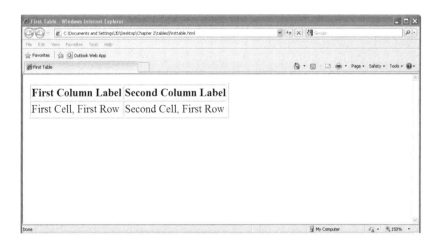

Figure 2.22 Table in a browser.

We can add some more information and enlarge our table. First of all, we could center it. If we were also to make it three columns and add another row, and include some cell padding around the content within each cell or table division, we would have the following:

```
<table border="2" cellpadding="5" align="center">

<tr>
<th>First Column Label</th>
<th>Second Column Label</th>
<th>Third Column Label</th>
</tr>

<tr>
<td>First Cell, First Row</td>
<td>Second Cell, First Row</td>
<td>Third Cell, First Row</td>
</tr>

<tr>
<td>First Cell, Second Row</td>
<td>Second Cell, Second Row</td>
<td>Third Cell, Second Row</td>
</tr>

</table>
```

This is the table with the additional content:

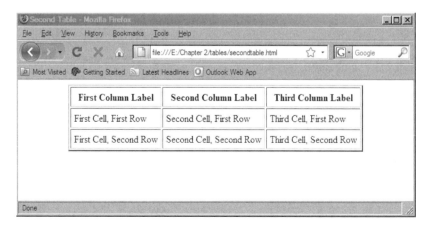

Figure 2.23 Three-column table in a browser.

Review the differences in the HTML code to see how we centered the table and added the padding. We have added "align="center"" for centering the table and we put some space around the text within each cell to make it easier to read with the "cellpadding="5"" designation, which puts 5 pixels between the cell borders and the letters. We also added an additional table row, table column, and table heading. Note how the heading row and two regular rows are designated by three separate sets of <tr></tr> tags.

We can also use color to indicate the width and height of cells or table divisions, and eliminate the border. This leads to a cleaner and simpler look that some designers prefer. To do this, we set the table border to 0 pixels, and do the same for "cellspacing." If we did not set cellspacing in this way, we would have some lines around each cell. Cellspacing is the spacing between cells, and if it is not designated the default setting for most browsers is 2 pixels.

In the example below we are also setting the width to 60% of the Web site area and using some inline styling with a style tag:

```
<table cellpadding="5" cellspacing="0" width="60%" style="font-
size: 10pt; font-family:'Arial', 'sans serif' ">
<tr>
<th colspan="2" align="center" style="background-color:#8FBC8F;
color:#ffffff">Species/Gender</th>
<th align="center" style="background-color:#b0c4de;
color:#ffffff">Number</th>
</tr>

<tr>
<td rowspan="2" valign="top" style="background-color:#e6e6e6">
Males</td>
<td style="background-color:#f5f5f5">Bluejays</td>
<td align="center" style="background-color:#f5f5f5">48</td>
</tr>

<tr>
<td style="background-color:#dcdcdc">Robins</td>
<td align="center" style="background-color:#dcdcdc">31</td>
</tr>

<tr>
<td rowspan="2" valign="top" style="background-color:#d3d3d3">
Females</td>
```

```
<td style="background-color:#f5f5f5">Bluejays</td>
<td align="center" style="background-color:#f5f5f5">29</td>
</tr>

<tr>
<td style="background-color:#dcdcdc">Robins</td>
<td align="center" style="background-color:#dcdcdc">40</td>
</tr>
</table>
```

Below we can see the borderless look of the table. In each row, we have indicated the background color that allows us to see the area of each cell without borders. In the first row, which contains the headings, the font color has been changed from default black to white with the hexadecimal color code "#ffffff." Note that we run the same color across two adjacent cells in the row to indicate the "Gender" and the "Species" and create the effect of one contiguous cell:

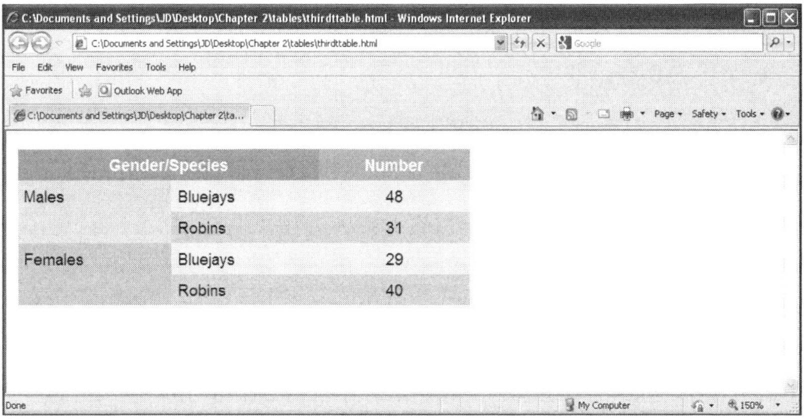

Figure 2.24 Borderless table with colspan and rowspan features in a browser.

For styling, we used the inline "style" tag within the opening table tag to designate the sans-serif Arial font, 10 points in height.

We also use the "colspan" and "rowspan" features of HTML that allow one division or cell to run across more than one column or row, respectively. For example, the row that designates the "Males" runs vertically across the two rows to the right of it for "Bluejays" and "Robins." The heading column for "Gender/Species" runs across the two columns below it.

Using tables for layout

In addition to presenting data such as the species, number, and gender of birds, tables can also be used to present larger bodies of information, such as text, links, and images. This was not the original intent when tables were first designed; as is often the case in the evolution of Web site production practices, designers began using tables for layout because it worked better with the available HTML tags.

Below is the HTML code for a simple table with the classic layout that we see on so many Web sites in use today; we will have a header, a left-side column used for navigation, an area for the main content of the page, and a footer, which is often used for copyrighting the Web's content. We will go into these general areas in more detail later, when we discuss CSS and layout, as the naming of these areas is a key part of the process of using CSS for layout. What is key here is that we are not using tables for data; instead, in the example below entitled "tableslayoutbirkerts.html," we are using them for text and links.

```
<body>
<table border="0px" cellspacing="0px" cellpadding="10px"
width="1000 px">

<tr>
<td colspan="2" style="background: #aababd" >Header</td>
</tr>

<tr>
<td style="background-color: #ebf5dc; width: 200px; height:
200px" valign="top">
<a href="tableslayoutbirkerts.html">Sven Birkerts</a><br />
<a href="tableslayoutbolter.html">J. David Bolter</a><br />
<a href="tableslayoutbernerslee.html">Tim Berners-Lee</a>

</td>

<td style="background-color: #ececec; width: 800px; height:
200px">

<p><h3>The Medium We Use for Writing</h3></p>

<p>The medium we read the word on—whether it is paper,
screen, or rock—has something to do with how we perceive it.
```

If a stone carver took so much time to carefully chisel out some
words in stone, we feel that these words are more important
than words that have been dashed off and instantly show up on
a screen. The writer using a pen or typewriter is "more likely to
test the phrasing on the ear, to edit mentally before committing
to paper" because it is just harder to write and/or type something out and then have to rewrite or retype it later (Birkerts
157). The traditional writer thinks carefully about what she is
writing, goes through all the word choices and phrasings in her
mind, not on the screen, and then puts them on paper. </p>
</td>
</tr>

<tr>
<td colspan="2" style="background: #aababd">Footer</td>
</tr>

</table>
</body>

As you go through this code, what is most important is identifying that
there are three table rows, and that, in the second row, there are two divisions
or cells that contain (1) the menu for navigation with some links, and (2) a
content area with some text. The first and third rows enclose the header and
the footer respectively. This is apparent in the screenshot below:

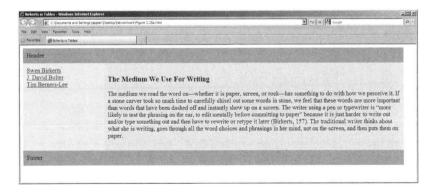

Figure 2.25 Simple table for classic layout in a browser.

As opposed to using a percentage for the width measurement—like we used in the table on bird statistics, which was set at 60% of the browser window—we are indicating that our table should run 1000 pixels across the top of the area available in a browser window. Because there is only one cell or <td> in each of the first and third rows, their widths will by default be the same width as the overall table, 1000 pixels. The width and the height of the two cells in the second row are 200 and 800 pixels, which total 1000 pixels. The height of both cells in the second row is the same: 200px.

We are using hexadecimal designated color codes such as "#aababd" to define the background area in each table cell, and the colspan tag to allow the first and third row to run across the two cells in the second row.

In the navigation bar we have three links, and if we were to add files other than the one above that characterizes some of the thinking of Sven Birkerts, we would keep the same basic structure for each one, with the same three links in place. In this way, every time we selected one of the links in the left-side column, we would land on a new HTML file that had the same layout look. What would change in each separate file would be the content in the second cell in the second row, with three different bodies of text character-izing the work of Sven Birkerts, J. David Bolter, or Tim Berners-Lee; we would have three separate table-based files, but they would be linked.

The code below is for a different table-based layout scheme, involving a table within a table:

```
<body>
<table border="0px" cellspacing="0px" cellpadding="10px"
width="1000px">

<tr>
<td colspan="2" style="background: #aababd">Header</td>
</tr>

<tr>

<td style="background-color: #ebf5dc; width: 200px; height:
200px" valign="top">
<a href="tableslayoutbirkerts.html">Sven Birkerts</a><br />
<a href="tableslayoutbolter.html">J. David Bolter</a><br />
<a href="tableslayoutbernerslee.html">Tim Berners-Lee</a>
</td>
```

```
<td style="background-color: #ececec; width: 800px; height:
200px">

<p><h3>Working at CERN</h3></p>

<table border="0px" width="200px" align="right">

<tr><td><img src="timbernerslee.jpg" width="200px" height="131px"
alt="Tim Berners-Lee"></td></tr>

<tr><td align="center">Tim Berners-Lee</td></tr>

</table>

<p>Many people have good ideas that are never accepted
because there is no community of people to begin applying the
ideas and making them the standard way of doing business.
Tim Berners-Lee was working at CERN, which is a particle physics
laboratory in Switzerland, and he saw his invention as a tool for
scientists to exchange ideas and support advanced research.
Berners-Lee's idea took hold because the people to whom HTML
was first introduced were technically able and saw that they
could actually use it; they understood, as Berners-Lee tells it,
"Much of the crucial information existed only in people's heads"
(Berners-Lee and Fischetti 9). This community of people saw
the value of exchanging information to advance scientific
discovery.</p>

</td>
</tr>

<tr>
<td colspan="2" style="background: #aababd">Footer</td>
</tr>

</table>
</body>
```

To place the image, we have embedded a separate table within the larger table construct, putting the opening and closing table tags in bold for easier identification. Here is the screenshot:

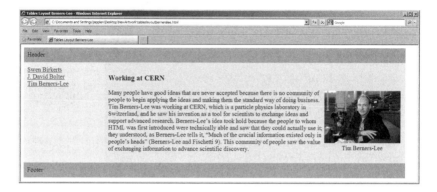

Figure 2.26 Embedded table in a table in a browser.

The embedded table sits within the second cell of the second row. The embedded table itself has two rows, one cell in each row. In the first row is the image, and in the second row is the caption, which we have aligned in the center. We have aligned the embedded table to the right of the second cell of the second row of the larger table. This technique is found in Castro (235) and provides a handy way of wrapping text around an image.

In the next variation of a table-based layout design, we are adding an additional cell to the second row but, unlike the previous variation, without an image or embedded table. Here is the HTML code:

```
</head>

<body>
<table border="0px" cellspacing="0px" cellpadding="10px"
width="1000px">

<tr>
<td colspan="3" style="background: #aababd" >Header</td>
</tr>

<tr>

<td style="background-color: #ebf5dc; width: 150px; height:
200px" valign="top">
<a href="tableslayoutbirkerts.html">Sven Birkerts</a><br />
```

```
<a href="tableslayoutbolter.html">J. David Bolter</a><br />
<a href="tableslayoutbernerslee.html">Tim Berners-Lee</a>
</td>

<td style="background-color: #ececec; width: 700px; height:
200px">

<p><h3>The Writing Space</h3></p>
<p>A "writing space" is a "material and visual field" that we com-
pose on and read from and it can take the form of such technolo-
gies as papyrus, paper, or a computer screen. J. David Bolter is
interested in how the changes in writing spaces have affected our
ability to think and communicate with each other: "Writing, even
writing on a computer screen, is a material practice, and it
becomes difficult for a culture to decide where thinking ends and
the materiality of writing begins, where the mind ends and where
the writing space begins" (12–3).</p>
</td>

<td style="background-color: #d8d8d8; width: 150px; height:
200px">
<p><b>Papyrus</b>: An ancient technology first seen in Egypt in
3000 B.C.</p>
</td>

</tr>

<tr>
<td colspan="3" style="background: #aababd">Footer</td>
</tr>

</table>

</body>

</html>
```

The third column variation allows us to add supporting information to the content, as the screenshot below illustrates. To do this, we added an additional cell to the second row.

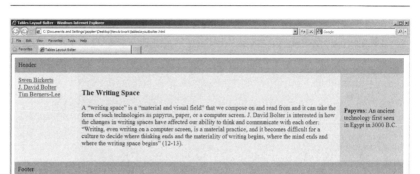

Figure 2.27 Three-column table layout in a browser.

Note that in the second row we have resized the cells so that all three of them fit within the 1000 pixel width: the first and third cells are 150 pixels each and the second cell, which constitutes the main content area, is 700 pixels. All are 200 pixels in height. We also had to change the colspan from 2 to 3 in the first and third row as we added an additional column.

The value of tables for layout

Tables are intuitive as they are reduced down to basic row and cell divisions that we can readily imagine. We have to realize that what we are doing is stacking one separate row on top of another to create the look of one contiguous table. Because they are easy to understand, tables are easier to manipulate and expand. They are easier to understand than CSS and are still used on many Web sites. Tables can be used to produce more complex designs than we have shown with the use of ingenious table-within-table embedding techniques.

The drawback is that some browsers do not render tables in the way you might intend, so always make sure that you test your tables on different browsers.

Exercises

1. Go through all of the tables in this section and manipulate their values.
2. For simple tables, add some more rows or columns, which might mean that you have to change the values for the colspan or rowspan elements. Also, manipulate the colors for the background of each cell in these tables. You can also change the style by manipulating the fonts and the position of the data within the cells.

3. Take any of the three table layout patterns we have discussed and embed the more conventional table, such as the example on birds, in the main content cell. Continue experimenting by placing this conventional table within a larger table with align center, right, or left designations.
4. Manipulate the width and height dimensions of each cell to produce a differently sized table.
5. Go online and view the source code of other Web sites that use tables for layout and analyze the ways others have employed <table>, <td>, and <tr> tags.

Works cited

Berners-Lee, Tim. "Long Live the Web." *Scientific American*. Dec. 2010: 80–85. Print.

Berners-Lee, Tim, and Mark Fischetti. *Weaving the Web: The Original Design and Ultimate Destiny of the World Wide Web by Its Inventor*. San Francisco, CA: HarperCollins, 1999. Print.

Birkerts, Sven. *The Gutenberg Elegies: The Fate of Reading in an Electronic Age*. Boston: Faber and Faber, 1994. Print.

Bolter, Jay David. *Writing Space: Computers, Hypertext, and the Remediation of Print*, 2nd ed. Mahwah, NJ: Erlbaum, 2001. Print.

Castro, Elizabeth. *HTML, XHTML & CSS*, 6th ed. Berkeley: Peachpit, 2007. Print.

Dickinson, Emily. *Poems by Emily Dickinson: Third Series*. Ed. Mabel Loomis Todd. Boston: Roberts Brothers, 1896. Print.

Gleick, James. "Patently Absurd." *New York Times Magazine. The New York Times - Breaking News, World News & Multimedia*, 12 March 2000. Web. 11 October 2010.

Chapter 3

Cascading style sheets

Chapter overview

Cascading Style Sheets (CSS) work in concert with HTML to more precisely define the style values we have been working with. In this chapter we will learn the basic syntax of this technology. We will apply CSS at the internal, inline, and external level and see how it can modify the style elements of entire Web sites much more efficiently than HTML can by itself. Additionally, we will devote a significant amount of time to describing how CSS can be used in layout design.

CSS

Cascading Style Sheets give us more style choices for our Web sites. Traditional HTML tags can specify some choices in fonts and colors, for example, but CSS expands on these choices.

CSS also allows us to control the style across a number of HTML files with greater ease. We can have one external style sheet that covers many separate HTML files for one large Web site reference, and when we change any of the style sheet specifications on the external style sheet, all the HTML files, because they refer to and are thus controlled by this one external style sheet, change accordingly. For instance, if you were a Web page architect and you just produced a very large Web site of some hundreds of separate yet interlinked HTML files for a large organization or corporate client, and the corporate client did not like the color combination of the font relative to the background color, you would not have to make changes in every HTML file. Instead, you would just change the font color and background color values on a single style sheet that all of your HTML files are linked to.

To be clear, external style sheets do not have any content. Instead, they govern how the content–the text and the images and other information–will look and be laid out on the Web site.

In this section, we will understand what "cascading" in CSS means and begin to see how CSS can be used to modify the paragraph tag, perhaps the most important tag for people who are interested in writing for the WWW.

Basic CSS syntax

The four essential parts of a style sheet are the selector, the declaration, the property, and the value. It is important to know these four terms and their definitions as they will be used throughout the rest of this book.

- **Selector**–A selector is the HTML element that you are designating. It could be a "p" for paragraph, "i" for italics, or any other element that serves to describe a conventional HTML style tag.
- **Declaration**–A declaration is the property and the value. These are enclosed within curly brackets, { }.
- **Property**–A property indicates to a browser the particular rule that is being applied. For instance, background-color and font-family are properties.
- **Value**–A value is specifically what the rule is set at. For example, it could be the actual name of the font, perhaps Arial or Helvetica, or the number of pixels that define the height of a font. It could be the actual color of the background, maybe blue.

For example, the following style rule for the background color on a Web page is indicated by the selector "body":

body {background-color: blue;}

The declaration is both the property and the value, and here the declaration consists of the property "background-color" and the value "blue." When the selector "body" is used in this context, the property is the "background-color" of the body and "blue" is the specific value. Note that between the property and the value we need to place a colon (:), and after this paired declaration we always need to put a semi-colon (;).

We can add as many CSS declarations as we want to each selector. For example, below is the "h3" heading that designates the font family and the color of the font:

h3 {font-family: arial; color: white;}

CSS allows for the multiple designations that more acutely specify style elements. Here we have two paired properties and values. The first is "font-family: arial," and the second is "color: white." Note the semi-colon that separates them and that they are both defining the h3 or heading 3 selector, an HTML element, which is designated <h3> when we see it in an HTML file.

We can also add multiple selectors. If we wanted all of our HTML headings, 1 through 6, to be Arial and white, we could add multiple selectors to the same declaration:

h1, h2, h3, h4, h5, h6 {font-family: arial; color: white;}

Note that the selectors are separated with commas.

Applying CSS

CSS can be applied to an HTML document at three different levels:

1. document or internal
2. inline
3. external

Document level or internal style sheet

At the document level, we put the styling information inside of the <style type="text/css"> and </style> tags, and this coding is placed between the <head> and </head> tags. These two <style> tags are shown in bold in the code below for ease of identification.

```
<!DOCTYPE html PUBLIC "-//W3C//DTD XHTML 1.0 Transitional//
EN"
"http://www.w3.org/TR/xhtml1/DTD/xhtml1-transitional.dtd">

<html xmlns="http://www.w3.org/1999/xhtml">

<head>

<title>CSS Practice</title>

<meta http-equiv="Content-Type" content="text/html;
charset=utf-8" />

<style type="text/css"> body {background-color:gray;}
h3 {font-family:Times New Roman; font-size:36px; color:white;}
</style>

</head>
```

```
<body>

<h3>Now is the time for all students to come to the aid of their
universities.</h3>

</body>

</html>
```

Here is how the above code looks in a browser. We did not indicate any of the style specifics in the <h3> or <body> tags. Instead, these are referenced in between the <style> tags, which in turn are between the <head> tags.

Figure 3.1 Document level or internal CSS.

With document level style tags, the <h3> element does not have its default appearance in the browser; the text is a different size and white. Similarly, the <body> element does not produce the default white background color; instead, it is gray.

Inline level CSS

Instead of placing style sheet information between the <head> tags so the entire HTML file is covered, we can use custom styled HTML elements that are different than the default tags. Here is an example using the same style for the <body> and <h3> tags that we used above in the document level CSS example, but at the inline level:

```
<!DOCTYPE html PUBLIC "-//W3C//DTD XHTML 1.0 Transitional//
EN"
"http://www.w3.org/TR/xhtml1/DTD/xhtml1-transitional.dtd">

<html xmlns="http://www.w3.org/1999/xhtml">

<head>

<title>Inline Style CSS</title>

<meta http-equiv="Content-Type" content="text/html;
charset=iso-8859-1" />

</head>

<body style="background-color:gray">

<h3 style="font-family:Times New Roman; font-size:36px;
color:white">Now is the time for all students to come to the aid
of their universities.</h3>

</body>

</html>
```

Start with a selector, in this case, <body> or <h3>, then follow with "style=" and the declarations (property and value) as needed, and put these in quotes. There is no need for curly brackets (i.e., {}) in inline syntax. As mentioned, we do not have to use any style coding elements between the <head> tags if we are using inline style elements. Save this file as "inlinestylecss.html." If you look at it in a browser, it will be identical to the example above, with a gray background and white text.

One of the strengths of inline syntax is that it overrides whatever other style sheets might be in place, either external or internal, and this allows us to fine tune our HTML/CSS code.

Producing a separate CSS file for an external style sheet

For an external style sheet, we first need to produce a simple Cascading Style Sheet employing the same code used above; we can then use it to style our HTML file. To do this we just take the code we have been working on and type it out in our editor as a text file:

body {background-color:blue;}
h3 {font-family:arial; color:white;}

Then we need to add a .css extension to it using our editor, saving it as a file entitled "firstcascadingstylesheet.css." Note this is different than the other files we have been saving, with the .html extension, but it is still a text file. We do not have to add any other coding apparatus to this CSS file, as we would in an HTML file–there is no need for <style> tags when we make external CSS files; it is just the two lines that you see here.

The next thing we need to do is add the following code within the head tags of each HTML file that is governed by our CSS file:

<link rel="stylesheet" type="text/css" href="**firstcascadingstylesheet. css**"/>

Note that we have specified "firstcascadingstylesheet.css" as the style sheet, following "href=." Review the entire HTML code below and see where the CSS reference (in bold for this example) has been added between the <head> tags:

```
<!DOCTYPE html PUBLIC "-//W3C//DTD XHTML 1.0 Transitional//EN"
"http://www.w3.org/TR/xhtml1/DTD/xhtml1-transitional.dtd">

<html xmlns="http://www.w3.org/1999/xhtml">

<head>

<title>CSS Practice</title>

<meta http-equiv="Content-Type" content="text/html;
charset=utf-8" />

<link rel="stylesheet" type="text/css" href="firstcascading
stylesheet.css" />

</head>

<body>

<h3>Now is the time for all students to come to the aid of their
universities.</h3>

</body>

</html>
```

Save this HTML file as "firstcssfile.html." By designating a style sheet in the code within the <head> tags, we are saying "Here is an HTML file governed by a Cascading Style Sheet file, which is a separate file that indicates that the background of the Web site when pulled up in a browser will be blue and the h3-sized text will be in an Arial font and white." Because we have specified values for the font such as its family and color, when we put our text between the <h3> tags, this is what we get, and here is how it would look:

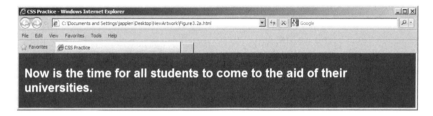

Figure 3.2 HTML file styled using an external CSS file, shown in a browser.

It is important to note that both the "firstcssfile.html" file and the "firstcascadingstylesheet.css" file are in the same folder. Pull up the "firstcssfile.html" file in your browser and you should see a blue screen with one line of text in white.

Had we produced an HTML file without designating the CSS extension file (firstcascadingstylesheet.css) between the <head> tags, we would get a default background color, font style, and color for the <h3> text, as we can see in the screenshot below:

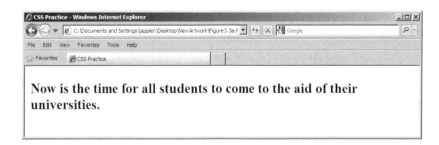

Figure 3.3 Default file without CSS.

One of the great things about CSS is that it allows us much more control over how browsers display our work. What we have done here is take a conventional HTML tag, <h3>, and redefine it so it has a different value.

We have already shown how to use the conventional HTML tag for paragraph, <p>, and this is one of the most common tags. By itself, it clearly separates paragraphs in a hypertext file. Using CSS, we can specify that the <p> means much more. In this example, it describes the font family, font size, line height, and color of the text:

```
p {font-family:Verdana, Arial, Helvetica, sans-serif; font-size: 12 px;
line-height: 1.5; color: #000000;}
```

Place this line of code in a new file and save it as "secondcascading-stylesheet.css." Again, this is the only line you need in the file; it is not an HTML file, so you do not need all of the HTML code apparatus described in the previous sections.

This line of code is telling us that we have indicated that Verdana should be the font style, and if the browser does not recognize Verdana, it should try Arial, and then Helvetica–not all fonts work on all browsers. The font size is 12 pixels, and the line height is 1.5 times this value, or 18 pixels, which is the distance between the baselines or bottom of the letters on successive rows. Line height establishes the "leading" between each line; "leading" is the word printers have been using for centuries to indicate the size of the gap they make between lines of type, originally by placing strips of lead between them. The color is standard black.

As we have changed the style sheet we are using, we will need to place the following line in between our <head> tags in the HTML file to indicate this:

```
<link rel="stylesheet" type="text/css" href="secondcascadingstylesheet.
css" />
```

Between the <body> and <p> tags, use the following text and save the HTML file as "secondcssfile.html":

```
<body>
<p>"Orality" is the word Walter Ong uses to characterize speech,
and "literacy" is the word he uses for written language. Ong not
only characterizes the nature of orality and literacy, more impor-
tantly he explains how both of these methods of communication
affect the way we think and understand our worlds.</p>
</body>
```

This is how our file will look:

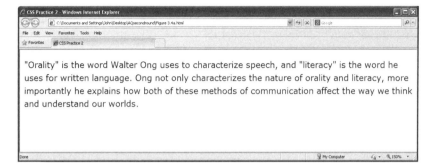

Figure 3.4 Second HTML file using external CSS.

The <p> tags are now doing much more than indicating where a paragraph begins or ends; they very acutely define how our text looks in a browser.

If we had constructed a very large Web site with many files and did not like the font choices, or wanted to change the font color or line height, all we would have to do is make some changes in the secondcascadingstylesheet.css file. This makes Web site construction and maintenance much more manageable, avoiding the need to go to our Web site files and make changes at the paragraph level. This is the greatest strength of an external CSS file.

As an exercise, make some changes in the CSS file, save them, and then review secondcssfile.html to see how it looks.

CSS classes

With the use of classes we can give a single HTML element a variety of possible looks on our Web sites. We do this by giving our tags more specific and unique selectors than plain HTML elements. These selectors allow for different styles and designate that they belong to a specific class.

For example, if we want to be able to produce a custom blockquote style that we will call "block," we can do this with the following code:

p.block {margin-left: 150px; margin-right: 150px; line-height: 1.2;}

The syntax for creating class styles requires that we place a dot or a period (.) before the name we have selected. For example, we use ".block" in the example above, then we add the already described CSS syntax for declarations, properties, and values between the curly brackets. In the example above, we are also adding a "p" to the class style as we want to make sure that the font style, font size, and color from the CSS paragraph description in the previous section are also the same in the block, thus we have "p.block."

Add the line above to the line we used in the previous section to describe paragraph, or "p," to make a new CSS file entitled "cssclassblock.css." Thus we have a new CSS file with the following two lines:

> p {**font-family**: Verdana, Arial, Helvetica, sans-serif; **font-size**: 12px; **line-height**: 1.5px; **color**: #000000;}
> p.block {**margin-left**: 150px; **margin-right**: 150px; **line-height**: 1.2;}

To understand the following three paragraphs, you will need to know the definitions of the selector, declaration, property, and value described in the "Basic CSS Syntax" section above.

Because there are no new declarations in the p.block selector for the font-family, font-size, and color properties already indicated in the p selector above it, these declarations will *cascade* down from the p selector to the p.block selector. This means that the p.block selector will still have the "Verdana, Arial, Helvetica, sans-serif" value, the "12px" value, and the "#000000" value for these respective properties. However, we have changed line-height from 1.5px to 1.2px in the p.block selector, so any HTML text that is coded using <p.block> HTML tags will have line heights of 1.2px, not 1.5px. The "newer" value overrides the "earlier" value.

Additionally, we have added some new code in p.block: there are some additional declarations in place that alter the left and right margins. The properties in these new declarations are "margin-left" and "margin-right."

The order of the respective selectors, p and p.block, is key here because the values *cascade* down from the first selector to the second. This is how CSS works. In summary, values for properties which appear in the first selector cascade down to the second selector. However, new values in the second selector for properties that also appear in the first will override the values in the first selector. Values for new properties which appear in the second selector only will also be applied.

```
<!DOCTYPE html PUBLIC "-//W3C//DTD XHTML 1.0 Transitional//
EN"
"http://www.w3.org/TR/xhtml1/DTD/xhtml1-transitional.dtd">

<html xmlns="http://www.w3.org/1999/xhtml">

<head>

<title>Using CSS Classes</title>
```

```
<meta http-equiv="Content-Type" content="text/html;
charset=utf-8" />

<link rel="stylesheet" type="text/css" href="cssclassblock.css" />

</head>

<body>

<p>A <i>fl&acirc;neur</i>, which is French for "stroller," finds
comfort in movement through a crowd:</p>

<p class="block">To the perfect spectator, the impassioned
observer, it is an immense joy to make his domicile amongst
numbers, amidst fluctuation and movement, amidst the fugitive
and the infinite .... To be away from home, and yet to feel at
home; to behold the world, to be in the midst of the world and
yet to remain hidden from the world&mdash;Baudelaire (qtd. in
Manovich 269).</p>

</body>

</html>
```

Note that we have referenced our CSS file, "cssclassblock.css," in the code between our <head> tags and that we have used a different syntax to mark the block passage: <p class="block">. We indicate that it is a class, followed by an equals sign, followed by the class name we have chosen, "block." We need to put the class name in quotes. Also, take notice of the special character for the "â" in *flâneur* (see the "Special Characters" section later in this chapter).

Here is how it would look in a browser:

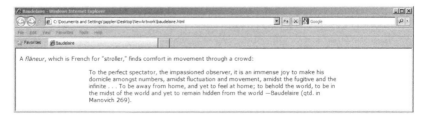

Figure 3.5 Using a CSS class for blockquote.

Note how the text-align, left-margin, and right-margin properties that are described in the p.block selector in the CSS code above differentiate the block from the lead sentence above it. If we had just used the stock HTML tag <blockquote> as opposed to the custom "p.block" class designation, this is how it would look:

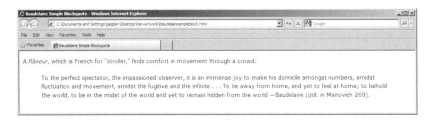

Figure 3.6 Default HTML blockquote.

Using classes in CSS allows us to further refine the look of our texts in a way that standard HTML tags cannot. As has been said before, if we had a large set of files on a large-scale Web site with many block elements in them and we wanted to change the look of these blocks, we would only have to make some changes in the CSS file and all of them would change. As an exercise, go to this book's Web site, download these files, and change the line height or color elements and see how this changes the style.

In this section we have demonstrated how CSS classes work to fine tune, among other things, the paragraph tag, <p>. We have described the following properties that can be associated with paragraphs in this section:

- line-height
- left-margin
- right-margin
- text-align
- font-family
- font-size
- color

For writers who want to control the look and placement of the words they write for their Web sites, knowing how to do this is important. These properties will be described in more detail in the sections that follow, and new properties associated with the <p> tag will also be covered.

CSS IDs

An ID selector is different than a class selector in that it can only be used to identify one element per HTML file that makes up a part of an overall Web site. In contrast, a class selector can identify numerous elements in the file.

ID selectors are most often used in layout design where we need to specify that specific parts of the HTML file, say the header, content area, footer, and left and right sidebars, are to be placed only once. We only need one header per HTML file, and one footer. In the example for classes above, we might have more than one <p class="block"> element as we might have text that contains a lot of extended quotations.

An ID selector is distinguished by the fact that it must begin with a pound or hash sign, #.

    #header {background-color:#FFD700;}

Here, we have designated the code for a CSS file that makes it clear that the background color of the header on a Web site should be #FFD700. In this case, the "header" is the rectangular space that runs along the top of a Web site, which usually contains the name of the Web site and some graphics.

It is a common practice to use IDs only for positioning elements relative to other elements in an HTML file, and we will follow this practice. In the "CSS and Layout" section of this book we will further illustrate how IDs can be incorporated into Web site design. Other IDs that we commonly see include "footer," "main-content," and "navigation" or "menu." As stated at the beginning of this section, we only use IDs for non-repeatable parts of Web files; for example, we would not have two footers or two headers.

CSS spans

Spans are a kind of class that we use when making an inline style change, usually in one sentence. Spans can be fine-tuned by using CSS so that the style of the text that is enclosed within the <span></span> tags will override any external or internal CSS code. For example, if we wanted to italicize a word or phrase in one sentence, we would start by placing the following line of code into the CSS file:

    .italic {font-style: italic;}

Add this line to the CSS file entitled "secondcascadingstylesheet.css" that contains one line for <p>, and save it as "thirdcascadingstylesheet.css."

After this, you need to use the span tag to alter the text that you want to place in italics. For example, use this line as the text in a new HTML file entitled "cssspanexercise.html":

    <p>Now is the time for <span class="italic">all</span> students to come
    to the aid of their universities.</p>

Make sure that you have included CSS file "href=" code between the <head> tags, using "thirdcascadingstylesheet.css." Here is how your new HTML file should look:

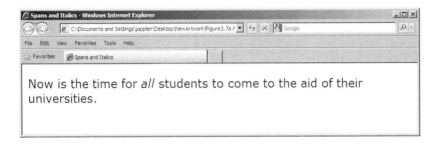

Figure 3.7   HTML file using a CSS span tag.

The "all" is in italics, and we have retained the font and other values from our previous CSS file containing the <p> designation, which describes font-family, font-size, line-height, and color for the text.

Using a span tag here is better than just putting <em></em> or <i></i> tags around the text we wanted italicized. If we wanted every italicized word to be bold in the entire document, we could change the code in the CSS file to:

.italic {font-style: italic; font-weight: bold;}

We would not have to do this in every place in a large Web site where there was a word, phrase, or line that needed to be italicized.

CSS divisions

If you want to style entire sections of your Web page, not just a word or phrase in a line, like when we use a span, or a body of words in a paragraph or blockquote, you should use a division. A division works in some ways like a paragraph tag, but you can style larger bodies of text with it. Division tags will break up a paragraph, so you cannot use them to alter just one part of a paragraph, as it will separate the text where you place the <div> tags into what looks like another paragraph.

You can include many elements, such as headings, paragraphs, and blockquotes, in a division. Divisions can be designed for different sections of large Web sites where different text sizes and style elements are required.

Make a CSS file using this text, and name it "divstylecss.css":

```
.opening {font-family: arial; font-size: 15pt;}
.main {font-family: arial; font-size: 13pt;}
h2 {font-family: arial; font-size: 16pt;}
h3 {font-family: arial; font-size: 14pt;}
```

Make a second file placing the following code between <body> </body> tags and name it "divstylehtml.html." Also make sure that you refer to the "divstylecss.css" file between the <head></head> tags in the HTML file.

```
<div class="opening">
<h2>Opening Heading</h2>
<p>This is an opening content sentence.</p>
</div>

<div class="main">
<h3>Main Content Heading</h3>
<p>This is a main content sentence.</p>
</div>
```

Note how the <div> tags are used to indicate two basic logical divisions, one entitled "opening," the other entitled "main." Below is how it would look in a browser:

These two divisions can be used to indicate logical structures commonly used in professional communication, with the font sizes for the different heading and text levels specified by reference to the appropriate division class in each set of division, or <div>, tags.

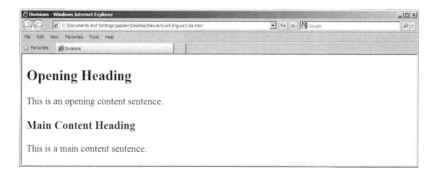

Figure 3.8 Common logical divisions of text.

Viewing CSS files

Studying others' code on Web sites is one of the best ways to improve our HTML and CSS skills. On Internet Explorer®, choose "View" and then "Source," and the HTML code will show up in your default editor. For Mozilla Firefox, choose "Tools," "Web Developer," and then "Page Source":

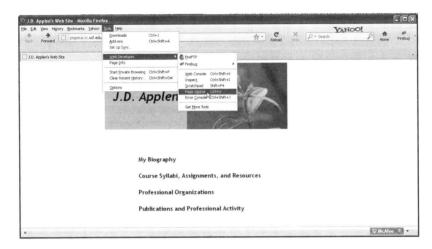

Figure 3.9 Viewing a Web page's HTML code in Mozilla Firefox.

This will reveal the Web page's HTML file coding:

Figure 3.10 Page source and CSS file link.

To get to the CSS code takes one further step. In the example shown, select "jdapplen.css" in the line that starts with "link," which designates the external CSS file. This reveals the CSS code in the default editor:

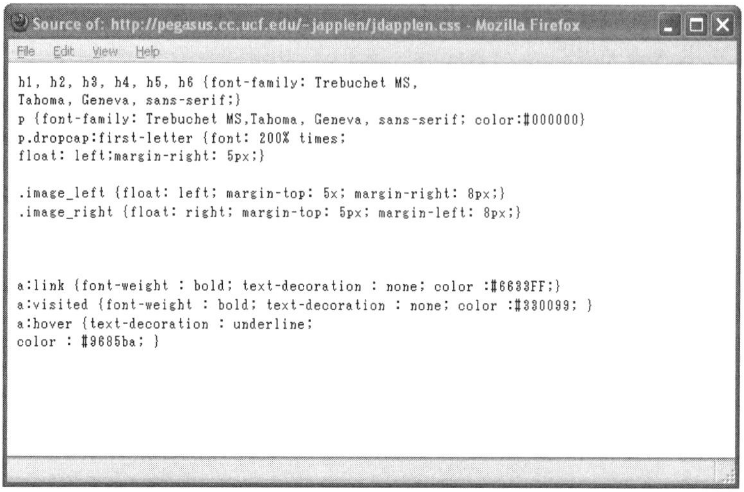


```
h1, h2, h3, h4, h5, h6 {font-family: Trebuchet MS,
Tahoma, Geneva, sans-serif;}
p {font-family: Trebuchet MS,Tahoma, Geneva, sans-serif; color:#000000}
p.dropcap:first-letter {font: 200% times;
float: left;margin-right: 5px;}

.image_left {float: left; margin-top: 5x; margin-right: 8px;}
.image_right {float: right; margin-top: 5px; margin-left: 8px;}



a:link {font-weight : bold; text-decoration : none; color :#6633FF;}
a:visited {font-weight : bold; text-decoration : none; color :#330099; }
a:hover {text-decoration : underline;
color : #9685ba; }
```

Figure 3.11 CSS file.

Often you can learn more from the external CSS file than from the HTML code that undergirds each page.

Some older browsers do not allow you just to select the code to get to the CSS file. Instead, you need to add the code to the opening URL of the Web site after a forward slash (/). In this instance, you would add to "pegasus.cc.ucf.edu/~japplen," the opening URL, "jdapplen.css," the designated CSS file, to get "pegasus.cc.ucf.edu/~japplen/jdapplen.css." Type this into the browser's dialogue box and the CSS file will be revealed.

CSS and layout

This section will illustrate how CSS can be used for layout design by employing it with (1) floats and layers of division elements, (2) absolute positioning, and (3) relative positioning. We will apply these to the classic vertical and horizontal menu layout designs that the vast majority of all Web sites are based on. While perhaps a little less intuitive than CSS for basic styles such as fonts, being able to use CSS for layout is an important skill, as it has become the standard format for today's Web sites.

CSS division elements for layout

We have already discussed divs when we showed how they could be used to more easily style a section of a Web page with a number of HTML elements.

We showed how both headers and font families could be styled within one division.

CSS divisions are more commonly used to describe where different sections of a Web page reside relative to one another: the layout. When you look at a Web site and see all of the different areas, such as menus containing sets of links, the Web site title, the content area(s), and the footer, you are looking at different division elements. They are block-level elements, or elements that take up or demarcate an area on a page. In Chapter 2 we used tables to set up these basic block-level elements.

The basic parts of the classic-looking Web site are:

- **Container**–We use "containers" or "wrappers" to enable us to control the layout of our Web sites. A container is a div element that includes all the other div elements that produce your basic layout structure. It describes the width of the Web site within the browser window.

  The container does not include any content, such as images, links, or text, only other division elements such as navigation bars, content areas, or footers. A container can be thought of as a box with a lot of other carefully packed boxes in it, and these smaller boxes have the content inside of them.

- **Header**–This division element runs across the top of the Web site and it is most often the space where Web designers present an image, a title, or both. Sometimes this is called a "banner."

- **Navigation**–This is usually either a left-side vertical column or a horizontal column running across the content divisions. It can also be described as the "menu," "nav," or "navbar," and contains links.

- **Content**–Generally, this is one or two columns where you put the major content, the images and the text, of your Web site. Sometimes content areas are given names such as "column-1," "column-2," or "main-content." However they are referred to, they are the areas that hold the bulk of a Web site's content.

- **Footer**–This block-level element resides at the bottom of your Web site and runs all the way across it. Footers are generally reserved for information such as copyright designation.

When we use CSS for layout, we are layering content or block-level elements such as these on top of each other. We start with the top or horizontal layer, or header.

Floating elements

One of the best ways to position block-level elements is by "floating" them. When we use the float property, we are asking that certain divisions "float" to the left or right side of the container. Floats are often used when we position images relative to text. Images are also block-level elements.

In the example below, we have some text and an image presented without using the float property:

```
<body>

<img src="timbernerslee.jpg">

<p>Many people have good ideas that are never accepted
because there is not a community of people to begin applying
an idea and making it the standard way of doing business.
Berners-Lee was working at CERN at the time, which is a particle
physics laboratory in Switzerland, and he saw his invention as a
tool .... </p>

</body>
```

As you can see in the following screenshot, if we just reference the image source and the text in sequential order, the text starts directly below the image:

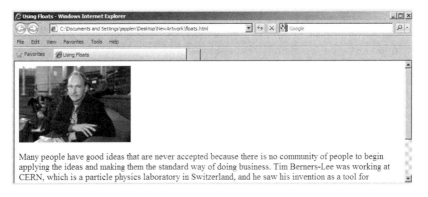

Figure 3.12 Image and text without float property.

In the following code, we use "float:left" to position the image relative to the text. In this circumstance, this is done with inline code, not CSS:

```
<body>

<img src="timbernerslee.jpg" alt="Tim Berners-Lee"
style="float:left; margin: 0px 15px 2px 0px">

<p>Many people have good ideas that are never accepted
because there is not a community of people to begin applying
```

an idea and making it the standard way of doing business. Berners-Lee was working at CERN at the time, which is a particle physics laboratory in Switzerland, and he saw his invention as a tool …. </p>

</body>

Below we can see how the image floats to the left of the first lines of text:

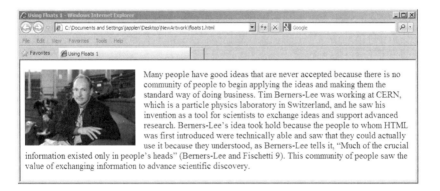

Figure 3.13 Image floated to left.

As you can see, we have floated the image to the left and the text appears on the right, as opposed to starting below the image. Contrast this to the way we used tables to encapsulate both an image and a caption in the tables and layout section. We have also added some margin spacing to the right of and below the image so the text and the image do not crowd each other. We did not set any margins at the top or to the left of the image; these margins were set to 0px.

Floats, divisions, IDs, and CSS

We floated an image in the previous section, but did not use CSS. Now we are going to float other block-level elements relative to one another–elements such as headers, footers, main content areas, and menus–and extend the use of floats with CSS. Review the CSS code that follows, which is saved in a file entitled "2-column.css":

```css
#container {
width: 1000px;
background-color: #ffffff;
margin: 0px auto;
}

#header {
width: 1000px;
text-align: center;
background-color: #cccc99;
}

#menu {
width: 150px;
height: 500px;
float: left;
}

#main-content {
width: 850px;
height: 500px;
float: left;
}

#footer {
width: 1000px;
clear: both;
background-color: #cccc99;
text-align: center;
margin: 0px auto;
}
```

Here is some HTML code which references the "2-column.css" file:

```html
<!DOCTYPE html PUBLIC "-//W3C//DTD XHTML 1.0 Transitional//
EN" "http://www.w3.org/TR/xhtml1/DTD/xhtml1-transitional.
dtd">

<html xmlns="http://www.w3.org/1999/xhtml">

<head>
```

```
<title>Left Menu 2 Column Header and Footer Layout</title>
<meta http-equiv="Content-Type" content="text/html;
charset=iso-8859-1" />
<link rel="stylesheet" type="text/css" href="2-column.css"/>

</head>

<body>

<div id="container">
<div id="header"><h1>Header, Menu, Main Content, and Footer
Layout</h1></div>

<div id="menu">
<ul>
<li><a href="home.html">Home</a></li>
<li><a href="bolter.html">Bolter</a></li>
<li><a href="birkerts.html">Birkerts</a></li>
<li><a href="berners-lee.html">Berners-Lee</a></li>
</ul>
</div>

<div id="main-content">

<p>The medium we read the word on–whether it is paper, screen,
or rock–has something to do with how we perceive it. If a stone
carver took so much time to carefully chisel out some words in
stone, we feel that these words are more important than words
that might have been dashed off and appear on a screen. ...
Regarding printed manuscripts, writers would be less likely to
write out something quickly. Instead, Birkerts believes that the
"writer was more likely to test the phrasing on the ear, to edit
mentally before committing to paper" because it is just harder to
write and/or type something out and then have to rewrite or
retype it later (157). The traditional writer thinks carefully about
what she is writing, goes through all the word choices and
phrasings in her mind, not on the screen, and then puts them on
paper. </p>

<p>However, perhaps Birkerts' ideal writer, who carefully thinks
through every word choice, could do the same on a computer;
the mental process of this writer is just facilitated with greater
ease–instead of trying to conjure up the correct words in our
```

heads, we do it on the screen. When we get it right, we stop and save it. Then we print it. Do the words on the page that eventually come from this process have any less meaning, power, or gravitas than words on paper or stone? Imagine reading the above passage by Birkerts after it had been carved in stone. Would you attribute a different meaning to it? What about if you read it from an ebook on a computer screen? </p>

```
</div>
<div id="footer"><h5>&copy 2013 Taylor & Francis</h5></div>
</div>

</body>

</html>
```

The # placed in front of the selectors in the CSS file designate them as "ID" selectors. For example, we have "#menu," not "menu." An ID selector is distinct from some of the other selectors we have been using, such as "h3" or "p," in that it can only be applied to one element in each HTML page or file. We will only have one container, header, footer, menu, or main-content selector in each HTML file, but we may have multiple "p" or "h3" selectors.

In the attendant HTML file, we need to designate each HTML tag for an ID selector with some additional text. For example, to indicate the menu, we use <div id="menu">.

Go through the <div id> designations in the example HTML file and note how they reference their respective selectors in the attendant CSS file. For example, we have <div id="main-content"> in the HTML file to reference the "#main-content" selector in the CSS file, which determines that the width of this block-level element is 850px, the height is 500px, and that it is floated to the left of the container. The code "margin:0px auto;" for both the container and the footer makes their left and right margins equal, thus centering these block-level elements. Because the container is centered, everything else inside it, everything that it "contains," is also centered. Go through the <div id> designations for the header, menu, and footer and note how their values are defined in the CSS file.

If we write "float: left," we are placing or floating the particular division to the left of the container or wrapper. Another way of thinking about "float: left" is that other division elements that are listed after an element that has been designated "float: left" will flow to the right of it (unless the divisions that are listed after have a "clear" instruction, which we discuss later). In the example, because we float the menu column to the left it will fit into the same layer as the main content column, which is also floated to the left. The widths of these block-level divisions, 150px and 850px, add up to 1000px, thus they fit within the width of the container, or wrapper, which is also 1000px.

If we write "float: right," we are placing the division element to the right of the container and the divisions that follow will move to the left of it, the converse of the "float: left" instruction. Note: If we do not indicate "float: left" or "float: right" in a division element that follows a division already floated to the left or right, some or all of the content could appear beneath the previous division, as opposed to the left or the right. You can see how this looks if you remove the "float: left" from the designation for main-content in the CSS file in the example.

The "clear" property in CSS allows you to override previous float properties. Using "clear: both" instructs the browser to move the division element below the divisions that came before it. Also, you can use "clear: left" or "clear: right" to clear left or right float commands listed earlier.

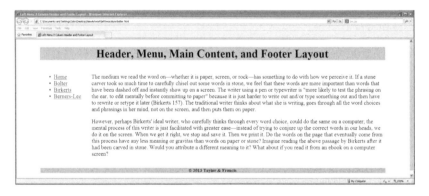

Figure 3.14 HTML file laid out using a two-column CSS file.

In the example, both the menu and the main content divisions are floating to the left of the container, but because the menu division, <div id="menu">, comes before the content division in the HTML file, it will float to the left first, followed by the content division. This example demonstrates the concept of layers. The footer division does not float to the left, but goes below what has come before it. If we removed the code "clear: both" from the CSS file, the footer would move up below the content division.

As an exercise, compare the code for the first example in the section on "Using Tables for Layout" in Chapter 2 to this last example. Both bodies of code–for tables and for CSS with relative positioning–produce essentially the same Web site structure, with a header, two columns, and a footer.

Note that in this two-column design, we needed to make the content area and the menu area the same color so that the menu area seems to be embedded within the content area. If we gave the areas different colors, the menu area would appear to stop just below the last link.

When we set the height of the menu and the content areas to be the same—in this case, 500 pixels—we can count on a consistent appearance as we select different links, but if we do not have very much content for one of the files, there will be a large space between where the text ends and the footer begins. If we did not set the height at all in the CSS code, the screen would adjust by default to enclose whatever text or other information we put in it. To understand how this plays out, reduce the height of one of these two divisions, the menu or main content, and give one a different background color from the other. In some of the layout architectures illustrated later in this book we will use coding such as "height: auto;" to automatically adjust for varying amounts of content in the main content division.

Using the same CSS style sheet, we can change the HTML and move the menu to the right side of the main content area:

```
<body>
<div id="container">

<div id="header"><h1>Header, Menu, Main Content, and Footer
Layout</h1></div>

<div id="menu">
<ul>
<li><a href="home.html">Home</a></li>
<li><a href="bolter.html">Bolter</a></li>
<li><a href="birkerts.html">Birkerts</a></li>
<li><a href="berners-lee.html">Berners-Lee</a></li>
</ul>
</div>

<div id="main-content">
<p>The medium we read the word on&mdash;whether it is
paper, screen, or rock&mdash;has something to do with how
we perceive it. ... </p>
</div>

<div id="footer"><h5>&copy;2013 Taylor & Francis</h5></div>

</div>
</body>
```

The graphic shows the menu on the right:

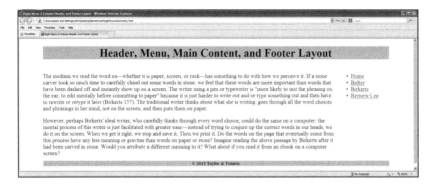

Figure 3.15 HTML for a two-column CSS file, with menu on right.

The difference between the left and right vertical menu constructions is where the menu ID coding has been placed in the HTML. Before, it was "above" the two paragraphs of main content.

Basic horizontal menu bar

Horizontal menu bars take a little more work. To make a basic horizontal bar, we would have to change some of the code in the CSS file to the code shown below, naming the file, say, "simplehorizontalmenu.css." The container, header, and footer selectors remain the same as in "2-column.css," but we are changing and adding to the menu and main-content selectors:

```
#menu {
width: 1000px;
float: left;
}

#menu li {
display: inline;
list-style: none;
padding: 0px 40px 0px 0px;
}

#main-content {
width: 1000px;
height: 500px;
clear: left;
}
```

First, the menu selector does not indicate 150 pixels in width (as it did in the vertical menu version) because we are going from a column to a horizontal navigation or menu bar. Instead, it is 1000 pixels in width and the menu bar runs the length of the container.

We are adding a "menu li" selector and placing within it the "display: inline" declaration to indicate that the links should not stack up on one another like in a vertical navigation bar. Instead, they will move across the menu bar horizontally and in sequence from left to right. We are also employing the "list-style: none" declaration to remove the bullets we always find in default unnumbered lists.

In the main-content selector, we have changed the declaration to "clear: left" so we can fit this block-level element directly below; this overrides the "float: left" declaration in the code above it. Previously the main content area was to the right of the menu column, both in the same layer. We could also have used "clear: both," which clears float: left and float: right declarations.

There is no need to change the HTML code at all, except to designate the new external CSS file, "simplehorizontalmenu.css." With these changes, we get the following layout:

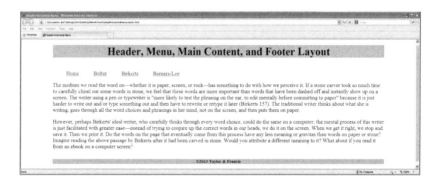

Figure 3.16 HTML file for a CSS file with horizontal menu.

From this exercise it is apparent that changing the external CSS file can change the layout and we do not have to alter the HTML file at all; we are using the same HTML file that we used for the left-side, vertical menu layout.

In the padding declaration in the "menu li" selector, we have designated that 40 pixels of space are added to the right of each link. If we added more links and needed to add to or reduce this variable to adjust the Web site's look, altering this number would be one way to do this.

CSS horizontal menu with tabs

Here is how to make a horizontal version of a single-column Web site with tabs. To get the tab look for our links in a horizontal arrangement, we can use the following CSS, putting it in a file named "csslayouthorizontal.css":

```
#container {
width: 1000px;
background-color: #ffffff;
margin: 0 auto;
}

#header {
text-align: center;
background-color: #cccc99;
padding: 15px;
}

#menu {
float: left;
width: 1000px;
background-color: #d3d3d3;
}

#menu ul {
margin: 0;
padding: 0;
}

#menu ul li {
list-style: none;
display: inline;
}

#menu li a {
float: left;
padding: 5px 20px 5px 20px;
border-right: 1px solid #000000;
}

#main-content {
clear: left;
width: 900px;
```

```
    padding: 15px 0px 0px 25px;
    }

    #footer {
    width: 1000px;
    clear: left;
    background-color: #CCCC99;
    text-align: center;
    margin:0 auto;
    }

    a:link {font-weight: bold; text-decoration: none; color: #000000;}

    a:visited {font-weight: bold; text-decoration: none; color: #ffffff;}

    a:hover {font-weight: bold; text-decoration: underline;
    color: #9685ba;}
```

Here is how the menu selectors contribute to the horizontal tab bar look:

- **#menu**–This characterizes the color and width of the menu. It is floated to the left side of the container.
- **#menu ul**–This allows us to define the margin and padding space around the unordered list ("ul") of links. We are resetting the default margins here to 0 pixels.
- **#menu ul li**–Because we have specified "list-style: none" for each list item ("li"), the bullets found in default unordered lists will not be displayed.
- **#menu li a**–Here we are making each of the four links float to the left, thus they line up horizontally. If we did not keep this float: left in the code, the links would stack up on top of each other vertically. We are also setting the padding of each cell and putting in a solid black (#000000) vertical line, 1 pixel in width, to the right of each tab, thus giving the tab bar a more distinctive look.

Here is the HTML for this Web site, from the link to the CSS file onwards:

```
<link rel="stylesheet" type="text/css" href="csslayouthorizontal.
css"/>
</head>

<body>
<div id="container">
```

```
<div id="header"><h2>Header, Horizontal Menu, Main Content,
and Footer Layout</h2></div>

<div id="menu">
<ul>
<li><a href="home.html">Home</a></li>
<li><a href="bolter.html">Bolter</a></li>
<li><a href="birkerts.html">Birkerts</a></li>
<li><a href="berners-lee.html">Berners-Lee</a></li>
</ul>
</div>

<div id="main-content">
<p>The medium we read the word on&mdash;whether it is
paper, screen, or rock&mdash;has something to do with how we
perceive it. ... </p>

</div>

<div id="footer"><h5>&copy;2013 Taylor & Francis</h5></div>

</div>
</body>
```

Below is the display in the browser:

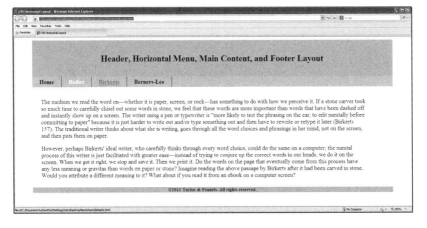

Figure 3.17 HTML file for a CSS file with horizontal tabbed menu.

Note how the CSS code for regular, visited, and hover links in the last three lines of the "csslayouthorizontal.css" file affect the links on the menu bar. Note also that the "border-right: 1px solid #000000" code separates each link with a black line.

CSS horizontal menu with two columns

Producing a Web site with two columns for content and a horizontal navigation bar requires that we add an additional container, a "content-container," that encapsulates the columns and ensures they line up correctly. Here is the HTML for this, again from the link to the CSS file onwards:

```
<link rel="stylesheet" type="text/css" href="csslayouthorizontal
twocolumn.css"/>

</head>

<body>
<div id="container">

<div id="header"><h2>Header, Horizontal Menu, Main Content,
Sidebar, and Footer Layout</h2></div>

<div id="menu">
<ul>
<li><a href="home.html">Home</a></li>
<li><a href="bolter.html">Bolter</a></li>
<li><a href="birkerts.html">Birkerts</a></li>
<li><a href="berners-lee.html">Berners-Lee</a></li>
</ul>
</div>

<div id="content-container">

<div id="main-content">

<h3>The Writing Space</h3>

<p>A "writing space" is a "material and visual field" that we com-
pose on and read from and it can take the form of such technolo-
gies as papyrus, paper, or a computer screen. J. David Bolter is
```

interested in how the changes in writing spaces have affected our ability to think and communicate with each other: "Writing, even writing on a computer screen, is a material practice, and it becomes difficult for a culture to decide where thinking ends and the materiality of writing begins, where the mind ends and where the writing space begins" (12–13).</p>

</div>

<div id="rightsidebar">

<p><b>Papyrus</b>: An ancient technology first seen in Egypt in 3000 B.C.</p>

</div>

</div>

<div id="footer"><h5>© 2013 Taylor & Francis</h5></div>

</div>

</body>

</html>

The encapsulated columns are the "main-content" and "rightsidebar" divisions; they are nested in the "content-container" division. Below is a screenshot of this file:

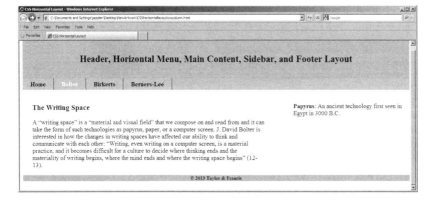

Figure 3.18 HTML file for a CSS file with a horizontal menu and two columns for content.

To get this layout, we need to replace the code in the previous CSS file for the "main-content" ID with the following, naming the new file "csslay-outhorizontaltwocolumn.css":

```
#content-container {
float:left;
width: 1000px;
}

#main-content {
float: left;
width: 600px;
padding: 15px 0px 0px 25px;
}

#rightsidebar {
float: right;
width: 300px;
padding: 15px 0px 0px 25px;
}
```

We have reduced the width of the "main-content" area to 600 pixels and added a "rightsidebar" area that is 300 pixels in width. Together they will still fit within the content container, 1000 pixels wide, that fits inside the first container ID, which encapsulates all of these divisions. The two columns only add up to 900 pixels, which gives us some extra horizontal space for the padding around the textual content.

Absolute positioning

Thus far in this discussion of CSS layout we have been employing the static positioning property with our use of document layers and floating commands. What this means is that there is a normal flow of the layers around each other that is specified by both their order in the HTML file and the float designations we give these divisions or block elements in the CSS file. The static property is the default, so no commands are needed.

Absolute positioning places the elements you are putting on a screen out of the normal flow of the document that we see in static positioning. For example, in static positioning we can count on one paragraph or image following another with default spacing and placement, and every paragraph and image we set down influences where the paragraphs and images that follow it will be set. In contrast, when we put an element in place using absolute positioning, what comes before this element does not affect its position, and what comes after it is unaffected by where we place it.

When we use absolute positioning, we are positioning block-level elements relative to the sides of the browser window, not to any wrapper or container. To place the block-level element, text, or image, we use the "top," "right," "bottom," or "left" properties to describe where they are positioned in relation to the sides of the browser window. We only need to use two of the four properties; providing coordinates or values for "top" and "left" or "bottom" and "right" properties is enough to position an element within a browser window.

In the example that follows, we start by placing the header, the first block-level element, at the upper left-hand corner of the browser window, which has coordinates 0 pixels from the top and 0 pixels from the left. Because we only need to label the starting point of the block-level element, we only need to label the top and left coordinates. Here we include both the CSS and the HTML in one internal file. Note that, as this is an internal CSS file, we are placing the CSS between the <head></head> tags. Because we are using IDs to describe our basic layout, in the CSS part of this file we start each block-level element with #:

```
<!DOCTYPE html PUBLIC "-//W3C//DTD XHTML 1.0 Transitional//
EN"
"http://www.w3.org/TR/xhtml1/DTD/xhtml1-transitional.dtd">
<html xmlns="http://www.w3.org/1999/xhtml">

<head>

<title>Two-Column, Absolute Positioning Web Site</title>
<meta http-equiv="Content-Type" content="text/html;
charset=utf-8" />

<style type="text/css">

#header {
background: #cccc99;
position: absolute;
top: 0px;
left: 0px;
width: 1000px;
height: 100px;
}
```

```
#menu {
background: #d7d7d7;
position: absolute;
top: 100px;
left: 0px;
width: 150px;
height: 450px;
}

#main-content {
background: #ffffff;
position: absolute;
top: 100px;
left: 150px;
width: 850px;
height: 450px;
}

#footer {
background: #cccc99;
position: absolute;
top: 550px;
left: 0px;
width: 1000px;
height: 50px;
}

</style>

</head>

<body>

<div id="header">Two-Column, Absolute Positioning Web Site</div>
<div id="menu">
<ul>
<li><a href="2-columnbolter.html">Bolter</a></li>
<li><a href="2-columnbirkerts.html">Birkerts</a></li>
</ul>
</div>

<div id="main-content">
```

```
<p>The medium we read the word on&mdash;whether it is
paper, screen, or rock&mdash;has something to do with how we
perceive it. If a stone carver took so much time to carefully chisel
out some words in stone, we feel that these words are more
important than words that have been dashed off and instantly
show up on a screen. The writer using a pen or typewriter is
"more likely to test the phrasing on the ear, to edit mentally
before committing to paper" because it is just harder to write
out and/or type something out and then have to rewrite or
retype it later (Birkerts 157). The traditional writer thinks about
what she is writing, goes through all the word choices and
phrasings in her mind, not on the screen, and then puts them on
paper. </p>
</div>

<div id="footer">Footer</div>

</body>
</html>
```

Here is this example in a browser:

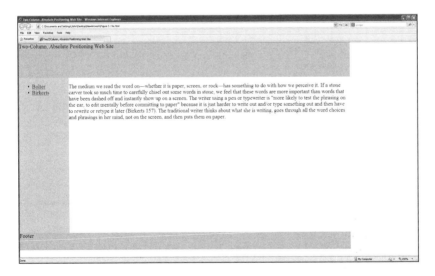

Figure 3.19 HTML file for a CSS file with absolute positioning.

Go through the code and make a note of where each block-level element begins. The header, as mentioned, begins in the upper left corner, 0 pixels from both the top and left. The menu begins at the top coordinate of 100 pixels, so it is below the header which is 100 pixels in height. The left side coordinate is 0 pixels, meaning the menu is still hugging the left side of the browser window. You can see all of the coordinates in the following screen capture, which uses the same layout but does not include text:

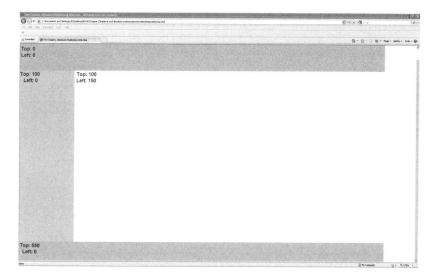

Figure 3.20 Two-column file showing coordinates for absolute positioning of block-level elements.

The "main-content" area has the absolute starting position of 100 pixels from the top and 150 pixels from the left. This is because it lies underneath the header which is 100 pixels in height and to the left of the navigation menu which is 150 pixels in width.

The footer is set at the absolute position of 550 pixels from the top as it lies beneath the header and menu areas, which have heights of 100 and 450 pixels respectively, totaling 550 pixels. Its absolute position from the top is 0 pixels, so it hugs the left side of the browser window.

If we do not get these numbers right, we can get overlap of these separate layout areas, or spaces between. For example, if the top of the menu was set at 125 pixels, not 100, and the top coordinate for the footer was 450 pixels, not 550, we would get the following look:

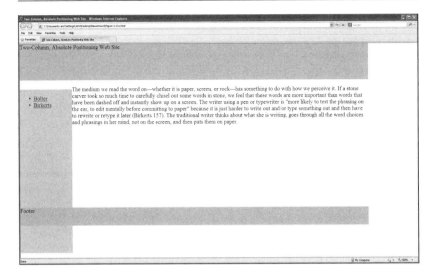

Figure 3.21 Two-column file with gap and overlap due to absolute positioning.

The point of this exercise is to show that, when using absolute coordinates, the different block-level sections will not float around one another as they do when we use static positioning. Instead, they assume their absolute positions and do not float.

Horizontal navigation using absolute positioning

If we changed the CSS code in the previous file to the following, we could make this a one-column absolute layout with a horizontal navigation bar, and we could also move the entire layout to the right and down by 50 pixels:

```
#header {
background: #cccc99;
position: absolute;
top: 50px;
left: 50px;
width: 1000px;
height: 100px;
}
```

```
#menu {
display: inline;
background: #d7d7d7;
position: absolute;
top: 150px;
left: 50px;
width: 1000px;
height: 50px;
}

#menu li {
display: inline;
list-style-type: none;
padding-right: 40px;
}

#main-content {
background: #ffffff;
position: absolute;
top: 200px;
left: 50px;
width: 1000px;
height: 450px;
}

#footer {
background: #cccc99;
position: absolute;
top: 650px;
left: 50px;
width: 1000px;
height: 50px;
}
```

Again, note that we do not have to float anything in this design; the elements are in absolute positions relative to the browser frame; we do not have any "float: left" or "float: right" code like we used in the container or wrapper layouts. Here is how it looks in a browser:

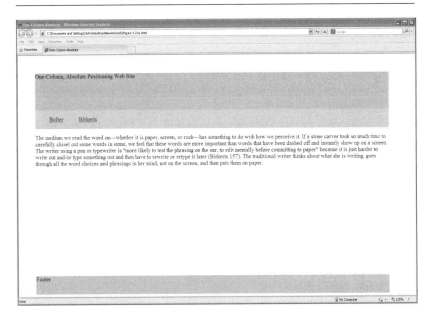

Figure 3.22 One-column file with horizontal menu and absolute positioning.

As you study this, identify the absolute "top" and "left" coordinates. The header is at "top: 50px" and "left: 50px," as opposed to 0 pixels for both coordinates in the previous example. When we do this, we have to change the absolute positioning coordinates for all of the other block-level elements. We also changed the dimensions of the "main-content" and "menu" elements. As an exercise, manipulate the top and left coordinates to see how this changes the layout.

Relative positioning

When we use relative positioning, we are positioning an object *relative* to where it would be in static positioning. This is in contrast to absolute positioning, when we place items specified distances from the sides of a browser window. As we did with absolute positioning, we use the "top," "right," "bottom," and "left" properties to place elements, but relative positioning works best with smaller elements such as text or images, rather than larger layout elements like menu, heading, main content, and footer areas.

When we use a positive value, such as 25 pixels, we are moving an object further from, for example, the "top" or "left" of its normal, default position. For instance, if we assign a value of "left: 25px," we are positioning the object 25 pixels further in from the left; with "right: 25px" we are positioning it another 25 pixels in from the right; similarly, "top: 20px" would move the object down by 20 pixels. When we assign such a value to an element, that element is placed relative to the preceding element by the assigned amount, say 20 pixels.

If we use the document-level, internal CSS code from between the <head></head> tags from the absolute one-column layout example and add the following HTML code, we can see how relative and static positioning elements work together:

```
<body>

<div id="header">One-Column, Relative and Absolute Positioning
Web Site</div>

<div id="menu">
<ul>
<li><a href="2-columnbolter.html">Bolter</a></li>
<li><a href="2-columnbirkerts.html">Birkerts</a></li>
<li><a href="2-columnberners-lee.html">Berners-Lee</a></li>
</ul>
</div>

<div id="main-content">

<h4 style="position: relative; left: 10px; top: 10px;">Tim Berners
-Lee</h4>

<img  src="timbernerslee.jpg"  width="200px"  height="131px"
alt="Tim Berners-Lee" style="float: left;">

<p>Many people have good ideas that are never accepted because
there is no community of people to begin applying the ideas and
making them the standard way of doing business. Tim Berners-
Lee was working at CERN, which is a particle physics laboratory in
Switzerland, and he saw his invention as a tool for scientists to
exchange ideas and support advanced research. Berners-Lee's
idea took hold because the people to whom HTML was first
introduced were technically able and saw that they could actually
use it; they understood, as Berners-Lee tells it, "Much of the
crucial information existed only in people's heads" (Berners-Lee
```

and Fischetti 9). This community of people saw the value of exchanging information to advance scientific discovery.</p>

</div>

<div id="footer">Footer</div>

</body>

Note that we have put some inline style positioning code in place for the <h4> tag: <h4 style="position: relative; left: 10px; top: 10px;">. We are using this code to move the heading "Tim Berners-Lee" 10 pixels further in from the left and 10 pixels further down from where the heading would be, in its default position. Here is how it looks:

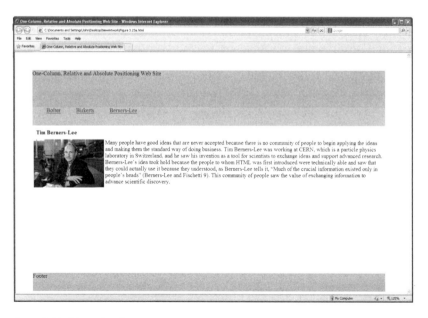

Figure 3.23 File with relative positioning.

Without this "position: relative" code, the heading would lie in its default position relative to the image and the text, 10 pixels higher and 10 pixels further to the left:

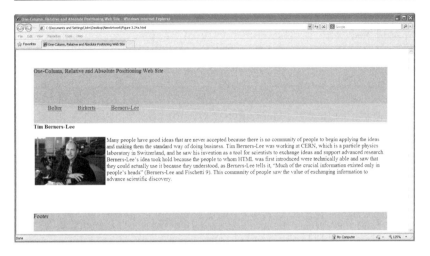

Figure 3.24 File with default positioning, and no relative positioning.

Regarding the flow of all the elements in a layout, when we apply a relative shift to one element, the others still work as if it were in its default position, the position it would be in with static positioning. Putting it another way, the other elements do not adjust their positions or float around the one element that has been moved. Thus it is wise to move elements by only a limited number of pixels. In this screenshot, the heading has been given "left: 25px" and "top: 25px" values, and has moved into the image:

Figure 3.25 File with relative positioning causing heading and image to overlap.

The flow of the image and text is based on the default position of the heading, not its actual position with the altered values of 25 pixels from the top and left; the image and text do not move down and to the right as the repositioned heading does, and this leads to the crowding we see here.

To move the inline style sheet code for the <h4> tag into the internal or document-level code between the <head></head> tags, we would change it to the following:

```
<head>
<title>One-Column, Relative and Absolute Positioning Web Site
</title>
<meta http-equiv="Content-Type" content="text/html;
charset=utf-8" />
<style type="text/css">
h4 {position:relative; left: 10px; top: 10px;}
</style>
</head>
```

This is also how the CSS code would look if it were in an external style sheet.

We could also adjust the image of Tim Berners-Lee by moving it down by 5 pixels using relative positioning. To do this, we would use this code:

```
<h4 style="position: relative; left: 10px; top: 10px;">Tim Berners-
Lee</h4>

<img src="timbernerslee.jpg" width="200px" height="131px"
alt="Tim Berners-Lee" style="float: left; position: relative; top:
5px;">

<p>Many people have good ideas that are never accepted
because there is no community of people to begin applying
the ideas and making them the standard way of doing business.
... </p>
```

Here is a screenshot showing this relative positioning change of 5 pixels down from the top, made using the "position: relative; top: 5px;" line of code:

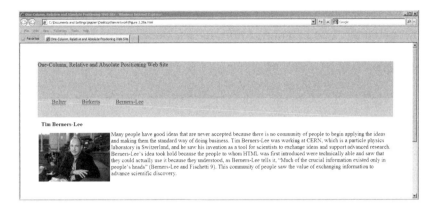

Figure 3.26 File with relative positioning of both heading and image.

Now the image of Tim Berners-Lee has been moved down 5 pixels, and it is more in line with the text line. Contrast it with the previous screenshots.

As illustrated, relative positioning works best for small adjustments. If one produced a large Web site using other layout technologies–technologies such as tables, CSS with a container or wrapper and floating elements, or absolute positioning–and we could not get an image or piece of text precisely in place, we could use relative positioning.

Alphabetic links

Producing a set of alphabetic links can be done with some simple HTML table structures. This first table contains one set of twenty-six horizontal links:

```
<body>

<h2>Alphabetical List of Links</h2>

<table>

<tr>
<td class="letterbox"><a href="alist.html">A</a></td>
<td class="letterbox"><a href="blist.html">B</a></td>
```

Note: *Twenty-two table divisions (<td></td>) covering letters C through X are elided here.*

```
<td class="letterbox"><a href="ylist.html">Y</a></td>
<td class="letterbox"><a href="zlist.html">Z</a></td>
</tr>

</table>

</body>
```

Here is how it looks in Internet Explorer®:

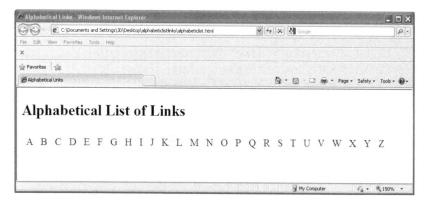

Figure 3.27 Horizontal alphabetic list using tables and CSS.

It is easy to see how this horizontal table could be set across a page and used as an organizing device.

The following is the HTML code for the "alist" file, or list of words that start with A. This also contains a table but in a different configuration: six rows of four table divisions each and one row of three table divisions. This second table includes all twenty-six letters plus an "All" link which returns us to the file with the first, horizontal table. Note that the table division with the "All" link uses the "colspan='2'" code so it spans two columns:

```
<body>

<table class="letterbox">

<tr>
<td class="letterbox"><a href="alist.html">A</a></td>
<td class="letterbox"><a href="blist.html">B</a></td>
<td class="letterbox"><a href="clist.html">C</a></td>
<td class="letterbox"><a href="dlist.html">D</a></td>
</tr>
```

Note: *The table rows for letters E to X are elided here.*

```
<tr>
<td class="letterbox"><a href="ylist.html">Y</a></td>
<td class="letterbox"><a href="zlist.html">Z</a></td>
<td class="letterbox" colspan="2"><a href="alphabeticlist.
html">All</a></td>
</tr>

</table>

<h1 class="alphalist">A</h1>

<p class="alphalist"><b>aardvark</b>&mdash;A medium-sized
mammal that we are more likely to see as the first word in the
dictionary than we are in the wild in Africa. "Aard" means "earth"
in the South African language and "vark" means "pig."</p>
<p class="alphalist"><b>aardwolf</b>&mdash;A small and furry
hyena. It follows "aardvark" in most dictionaries.</p>

</body>
```

In the following screenshot we have the "A list" definitions and the link structure now in a different table form:

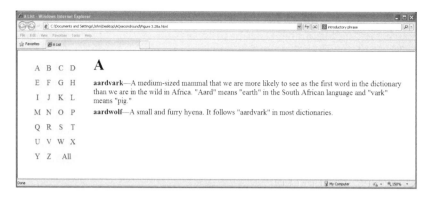

Figure 3.28 Alphabetic list on left sidebar.

This table-structure "A-list" HTML file allows for a workable placement of a group of links, set to one side, in this case the left side, of the screen. The "All" link takes the audience back to the horizontal set of links, and they can also go to any other indexed letter.

Both files use the following CSS, set up as an external link:

```
table.letterbox {margin: 10px 30px 10px 20px; float: left;}

td.letterbox {background-color: #f8f8f8; padding: 5px 5px 5px
5px; text-align: center;}

a {text-decoration: none;}

p.alphalist {margin: 5px 5px 5px 5px;}

h1.alphalist {margin: 10px 10px 10px 5px;}
```

Consider how this CSS contains all of the style and layout elements for the "A-list" HTML file. For instance, the "float: left" declaration for the "table.letterbox" selector sets the table to the left of the heading and paragraph elements. If we wanted to move the table to the right, all we would have to do is change the code to "float: right." This would mean that all of the twenty-six files for letters A–Z would have a different layout scheme because we altered some coding in the CSS file.

The "table.letterbox" and "td.letterbox" selectors only determine the specific styling for this table and its table divisions. If there were other tables on the Web site that used other selectors, such as "table" or "td," their styles would not be determined by these two selectors.

The "p.alphalist" selector is for the paragraphs containing the items in the A list, such as "aardvark." If there was another "p" selector that came before this selector and designated specific font values, such as the size, the paragraphs defined by the "p.alphalist" CSS selector would inherit these values. The same applies to the "h1.alphalist" selector.

To see how this works, produce the files above in an editor and then alter the CSS coding to the following and see how it changes the overall layout and style:

```
table.letterbox {margin: 10px 30px 10px 20px; float: right;}

td.letterbox {background-color: #f8f8f8; padding: 5px 5px 5px 5px; text-align: center;}

a {text-decoration: none;}

p {font-family: Calibri, sans-serif; font-size: 1.0em;}

p.alphalist {margin: 5px 5px 5px 5px;}

h1 {font-family: Arial, sans-serif; font-size: 2.4em;}

h1.alphalist {margin: 10px 10px 10px 5px;}
```

Here is how it would look in Internet Explorer®:

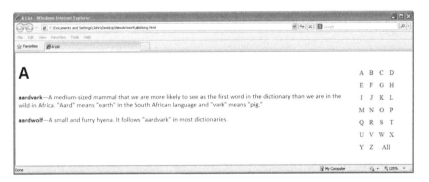

Figure 3.29 Alphabetic list on right sidebar and with altered paragraph styles.

Review the changes in the CSS file and consider how they have changed the layout of the "A-list" file.

CSS layout and breadcrumbs

You can give people a visual sense of your Web site's hierarchy with the use of simple breadcrumbs, which are a record of the links that lead to the one you are currently on. With breadcrumbs, patrons can navigate "backwards" or "up" to the link that records where they were previously. This is especially useful on Web sites with many links and deeper hierarchical structures.

The flowchart of a Web site hierarchy below reveals the distance of each link from the home page and the overall structure. One link away from the home page lie three separate "Tier 1" files, and below these are their respective "Tier 2" links. The word "tier" is used here to indicate the level each Web page is on and where it resides relative to the home page; Tier 1 pages are one link away, and Tier 2 pages are two links away. This designation, originating in the work of Ginny Redish, is discussed in more detail in Chapter 5 in the section on "Pathway or First-Tier Pages."

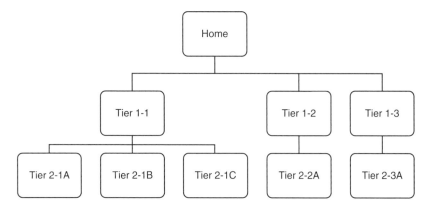

Figure 3.30 Web site hierarchy and tiers for breadcrumbs.

Many Web sites have a more complex hierarchy and a deeper structure than the one shown, but it will serve our basic purposes. If we were to go farther down the hierarchy than the second tier, we can imagine an expanding set of choices or links. Note that if we go to Tier 1-1, we have three choices: Tier 2-1A, Tier 2-1B, and Tier 2-1C.

The following CSS code is for a basic layout with a horizontal links navigation bar or menu that represents the first-tier links in the hierarchy:

```
#container {
width: 1000px;
background-color: #ffffff;
margin: 0 auto;
}

#header {
text-align: center;
background-color: #cccc99;
padding: 15px
}

#menu {
float: left;
width: 1000px;
background-color: #d3d3d3;
font-family: Verdana, Arial, Helvetica, sans-serif; color: #000000;
font-size: .8em;
}

#menu ul {
margin: 0;
padding: 0;
}

#menu ul li {
list-style: none;
}

#menu li a {
float: left;
padding: 5px 20px 5px 20px;
border-right: 1px solid #000000;
}

p.breadcrumbs {
height: 25px;
padding: 30px 20px 5px 20px;
font-family: Verdana, Arial, Helvetica, sans-serif; color: #000000;
font-size: .8em;
}

#main-content {
```

```
clear: left;
width: 900px;
padding: 15px 0px 0px 25px;
}

#footer {
width: 1000px;
clear: left;
background-color: #CCCC99;
text-align: center;
margin: 0 auto;
}

a:link {font-weight: bold; text-decoration: none; color: #000000;}
a:visited {font-weight: bold; text-decoration: none;
color: #3399ff;}
a:hover {font-weight: bold; text-decoration: underline;
color: #9685ba;}
```

The following is the HTML code for the home page, or "Home," using this CSS code, from the first <body> tag onwards. Make sure you refer to the CSS file between the <head></head> tags.

```
<body>

<div id="container">
<div id="header"><h2>Header, Horizontal Menu, Breadcrumbs
Menu, Main Content, and Footer Layout</h2>
</div>

<div id="menu">
<ul>
<li><a href="homebreadcrumbs.html">Home</a></li>
<li><a href="tier1-1breadcrumbs.html">Tier 1-1</a></li>
<li><a href="tier1-2breadcrumbs.html">Tier 1-2</a></li>
<li><a href="tier1-3breadcrumbs.html">Tier 1-3</a></li>
</ul>
</div>
```

```
<p class="breadcrumbs"></p>

<div id="main-content">

<p>Home Content</p>

</div>

</div>

<div id="footer"><h5>&copy; 2013 Taylor & Francis</h5></div>

</body>

</html>
```

The screenshot below represents what this layout would look like if we were on the home or index page of the Web site:

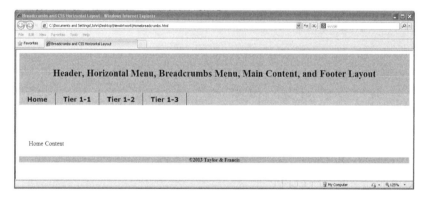

Figure 3.31 Home page content without breadcrumbs.

In this case, there is no breadcrumb content between the <p> tags in the HTML code, <p class="breadcrumbs"></p> (which are shown in bold above). Because of this, there are no breadcrumbs on the screen.

If we select the link "Tier 1-1," having already set up the file "tier1-1breadcrumbs.html," we go to the following screen:

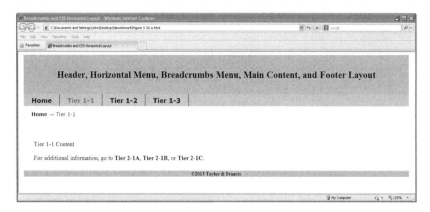

Figure 3.32 Tier 1-1 content with breadcrumbs.

On this page, we have some "content" and three additional link choices, all in the "main-content" area. This file is one link away from (or below) the home page, on the first "tier" or level, along with the Tier 1-2 and 1-3 files. To produce this Tier 1-1 page, the following HTML code is used to replace the code between the <p class></p> tags and the "main-content" div tags in the previous code; this is the code that we use for the Tier 1-1 link and that distinguishes this page from the home page:

```
<p class="breadcrumbs"><a href="homebreadcrumbs.
html">Home</a> &rarr; Tier 1-1</p>

<div id="main-content">
<p>Tier 1-1 Content</p>

<p>For additional information, go to <a href="tier2-1Abread-
crumbs.html">Tier 2-1A</a>, <a href="tier2-1Bbreadcrumbs.
html">Tier 2-1B</a>, or <a href="tier2-1Cbreadcrumbs.
html">Tier 2-1C</a>.</p>

</div>
```

Note the new code between the <p class> tags in this version. In the "home" version, the <p> tags were left in, but no breadcrumb content was included. Note that the code used for the arrow between the breadcrumbs here is "→".

In addition, there are three links that can be selected: Tier 2-1A, Tier 2-1B, Tier 2-1C. If we choose the link Tier 2-1A (again, having set up the appropriate file), we get to the following page:

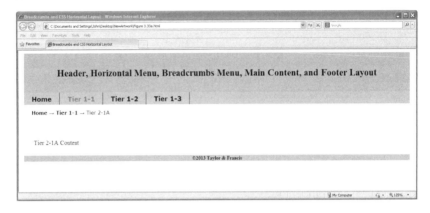

Figure 3.33 Tier 1-A content.

Note that the breadcrumbs have expanded to show that we are on Tier 2-1A, but have come to this page through Tier 1-1. It also shows us that we are two links in sequence away from the Home page. To get to this, we need to use the following code in our HTML version of Tier 2-1A:

```
<p class="breadcrumbs"><a href="homebreadcrumbs.
html">Home</a> &rarr; <a href="Tier1-1breadcrumbs.
html">Tier 1-1</a> &rarr; Tier 2-1A</p>

<div id="main-content">

<p>Tier 2-1A Content</p>

</div>
```

Review this code and note how we have to build in the breadcrumbs links. As an exercise, produce a Tier 1-2 or 1-3 page and some additional pages below it.

Breadcrumbs are best used on Web sites that are hierarchically organized and on which each successive tier can be logically understood to be grouped within the previous tier. For example, an appropriate grouping for the Web site of an online bookseller might be a "textbooks" link, and below this a "used textbooks" link.

Breadcrumbs should be considered a secondary navigation scheme, not the primary navigation scheme. The Tier 1-1, 1-2, and 1-3 links that are available in the first navigation line when first arriving at the Web site constitute the primary navigation scheme.

Breadcrumbs should be made into links so the user can easily retrace her steps to previous pages.

In these examples we have been using the code "→" for the arrow symbol signaling each new breadcrumb; we can also use ">" or ">" to produce a "greater than" sign, >.

Images

In many cases, images are an essential ingredient of a Web site, and they can reinforce its ethos or authority. However, images for the sake of images should not be used if they do not complement the text. Of course, there are instances in which images are the most important element on a Web site, for example on a professional photographer's site or for a museum curator who decides to present the most famous works in his gallery.

Image file formats

JPEGs and GIFs are the two most popular image formats in use on the Web today, and it is important to understand the differences between them when deciding which one to use.

GIF (Graphics Interchange Format) files are able to represent 256 colors and most often they are used for simpler images such as logos and line drawings, or any graphics that have larger areas of the same color. They are not recommended for photographs as they have a limited color palette and cannot capture the many gradations of color that photographs usually have. The advantage of GIFs is that they load faster. Use the .gif extension for GIFs.

JPEG files have the ability to display many thousands of colors and are thus well suited to use for photographs and other images with many colors and gradations. JPEG is an acronym for Joint Photographic Experts Group. The .jpg or .jpeg extension is used for JPEGs.

Compressing an image is like reducing it down to its bare essentials, with no redundant coding taking up space. When we compress a file, we reduce the overall amount of digital information that it contains, and because it is smaller it will load more quickly. However, when files are compressed there is the danger that reducing the information in the file will mean the quality of the image is compromised.

Images and HTML

When placing an image in a Web site file, you need to use the <img src> HTML tag:

    <img src="name of image file and extension"/>

In addition to this relative link to the jpeg file, which in this case is on the same hard drive as the HTML code and text, you can also place an image on a server and make an absolute link to it using the following code:

    <img src="http:// URL and name of image file and extension"/>

The code "img src" is short for "image source." Unlike most HTML coding, there is no need for a closing tag with the <img src> tag. However, we do need to add a forward slash, or /, when we close off this tag. As discussed, the extensions on image files are usually either ".gif" or ".jpg," and they are preceded by the name of the file, for example "purpleflower.jpg."

In addition, we need to put in the height and the width of an image, in our example 190 pixels and 300 pixels. This is because we need to create a space for the image in the layout of our page.

Finally, one of the best practices for Web designers is to include an "alt" attribute which provides an "alternative text" in case the image does not download or downloads slowly. Here is the alt tag integrated with the code for our image, which includes the height and width:

    <img src="purpleflower.jpg" height="190px" width="300px" alt="purple
    flower"/>

In an age when people generally have faster Internet connections than in the past, the alt text is not quite as important–images usually download with sufficient speed.

Here is the complete code for these HTML properties and some text to accompany the image:

```
<img src="purpleflower.jpg" height="190px" width="300px"
alt="purple flower"/>

<p>This is a flower in my back yard, and it looks like it is getting
ready to bloom. You never know what you will find close to where
you live; you just have to open your eyes and look. </p>
```

Below is how it looks in a browser. Note that the alt text "purple flower" appears when the pointer hovers over the image:

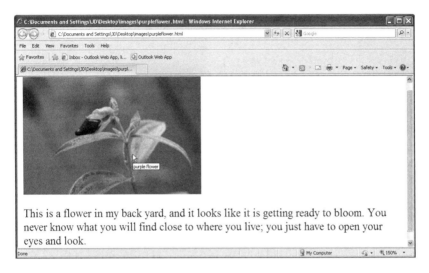

Figure 3.34 Image and text.

In this instance we get the image with the text lining up below it. However, using the "align" HTML attribute allows us to better position images and other objects relative to text. The values associated with the align tag are "left" and "right." The code for aligning the image in our example to the right is:

```
<img src="purpleflower.jpg" height="190px" width="300px"
align="right" alt="purple flower"/>
```

Here we have the image aligned to the right and text set to the left. A heading and some additional text has been added to this file to show how the text wraps around the right-aligned image:

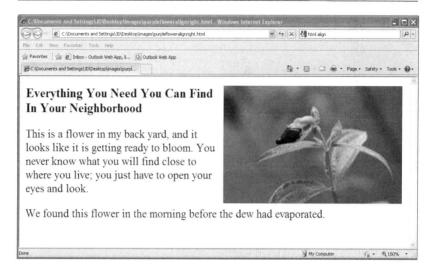

Figure 3.35 Image aligned to right of text.

In the section "CSS and Layout," we described how CSS and "float: left" and "float: right" declarations can be used to place images. It is important to know this because "align" is a deprecated tag, though it is still found on many Web sites and supported by today's browsers. Going forward, it is best to use float: left and float: right to position images, but it helps to know how the "align" tag works.

Image borders

Usually a borderless image makes a Web page look clean and uncluttered, but at times a border can help set the image apart from the rest of the layout elements and give it greater impact.

In this example, we add a border of four pixels:

```
<img src="purpleflower.jpg" height="190px" width="300px" alt="purple
flower" border="4px"/>
```

Below is how it looks in a browser with the border in its default black:

Figure 3.36 Image with border in default black.

The default black border distinguishes the image from the space around it.
A border is best used when the image has a similar color to the back-ground color of the body, as it makes the image more distinct:

Figure 3.37 Image with border against similar background color.

Without the border, it is difficult to see where the background ends and the image begins. Here is the code for this image, using a border value of 2 pixels, from the <head> tag onward:

```
<head>

<title>Purple Flower with Border</title>
<meta http-equiv="Content-Type" content="text/html;
charset=iso-8859-1"/>
<style type="text/css">
p, h3 {font-family: Arial, Verdana, Trebuchet, sans-serif;
color: white;}
body {background-color: gray;}
</style>

</head>

<body>

<img src="purpleflower.jpg" height="190px" width="300px"
align="right" alt="purple flower" border="2px"/>

<h3>Everything You Need You Can Find In Your Neighborhood
</h3>

<p>This is a flower in my back yard ... before the dew had evapo-
rated.</p>

</body>

</html>
```

As an exercise, you might alter this file by changing the border value to "0px" to see the effect of removing it.

We have added some CSS to the file that changes the "font-family" property and the color of the font. This code is placed between the <style> </style> tags, which in turn are encapsulated within the <head></head> tags for the file.

Image links

We see image links on Web sites all the time. Usually they are icons, such as a small box or rectangle with a picture of a simple house for "home," or a hand pointing in one direction or another to signify "return to previous page" or "move to next page." You select the image and it takes you to an absolute or relative link. You can also produce simple image links containing text, such as this "Home" link, using graphics editing software—even though it includes text, it is, at an HTML level, an image:

HOME

Figure 3.38 "Home" image link.

If you are going to use a graphic as an image link, you need to understand that the graphic is going to have to cue the reader so that she will have some sense of where it will lead. Image links of products work well in this way. If you select an image of a laptop computer on a Web site showing thirty images of different laptops, it is most likely understood that this image will lead you to further information about that particular laptop, perhaps even a specification sheet.

Other image links can be more playful. For example, this classic image might be used to suggest that if you select it, it leads to another screen that describes the actions people can take to solve a problem or engage in some kind of community activity that advances a cause:

Figure 3.39 Classic image link.

Similarly, an image of Tim Berners-Lee might suggest that, if you were to select it, you would go to the W3C Web site, as he is so clearly associated with that organization. The key point here is that the image should accurately suggest to the particular audience what lies beyond it if it is selected. Image links can liven up a page and act as visual shorthand that makes Web navigation more rewarding for the viewer.

An image link consists of an absolute or relative link and the image associated with it:

<a href="absoluteorrelativelink.html"><"image.jpg"></a>

Here is the code for a Tim Berners-Lee image link to the W3C Web site:

<a href="http://www.w3.org"><img src="timberners-lee.jpg"></a>

This is the Tim Berners-Lee graphic with the mouse-sensitive icon over the image:

Figure 3.40 Tim Berners-Lee image link.

If we select this image, we are led to the W3C Web site:

Figure 3.41 W3C absolute link destination.

Resizing a file

If you resize an image, for instance, and go from a larger image to a smaller image, then save it, you are restructuring the binary information that makes up the image. If you then take the smaller version of the image and make it larger again, you are further altering the code and the image quality will be degraded, with a loss of detail. It is best to take the original GIF or JPEG and use it to produce a larger or smaller image for a specific purpose. When you do this, always give the resized image a new name and keep a copy of the original image for further manipulations.

There are several reasons we include the height and width properties in the code for an image. One is that this creates a space matching the dimensions of the image which can be seen before the image completely downloads, and a viewer can see that an image will most probably be appearing, depending on the connection speed.

You can also use the height and width values to make a larger image fit into a smaller space or vice versa. When a larger image is made to fit into a smaller space, the quality of the image does not change, but fitting a smaller image into a larger space usually results in poorer image quality, with less detail. The height and width values assigned to an image can help you size it, but there is a cost involved as it will be less detailed.

Instead of putting an image into a space that is larger or smaller than its particular dimensions, it is good practice to resize it. This is especially true of larger images, which take longer to load. Resizing an image is an important skill, as archived images rarely meet the exact size requirements of your layout. This technique will be demonstrated in the next section.

Thumbnail image links

A thumbnail image link is a link from a smaller version of a graphic to a larger version. This works well if you want to present your viewer with one image or an array of images that will load relatively quickly, and also allow your viewers the choice of loading a larger image. Such links may be used for larger images that have a lot of detail, such as fine art, portraits, and maps or schematics.

To produce a thumbnail image link, you will need two images: a smaller one that takes up less digital space so it downloads faster and that also takes up less space on the screen, and the larger one of higher quality that you want to make available to your viewers.

Usually the original image that you start with is the larger version, and you will need to reduce its size using graphics software. To do this, pull up the original image in graphics editing software and change the values in the width and height fields, then save it with a different name that indicates it is the thumbnail version. This is shown here with Adobe™ Photoshop™, but other graphics editors can perform the same functions.

Here is the larger version of the Tim Berners-Lee image with the "Image Size" dialogue box:

Figure 3.42 Image size dialogue box and large image.

This "timbernerslee.jpg" file has a width of 2000 pixels and a height of 1312 pixels. To reduce this image, the width was changed by typing "200" into the width field, as shown in this screenshot:

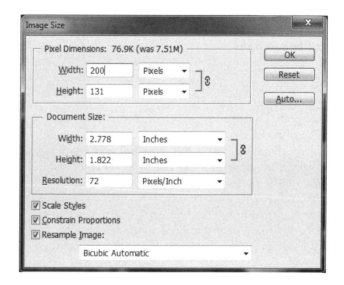

Figure 3.43 Image size dialogue box with reduced values and "Constrain Proportions" selected.

Because "Constrain Proportions" was selected, when "200" was typed in, the value in the height box was reduced to 131 pixels automatically, and the image retained its relative proportions. The smaller version is 10 percent of the original size.

After this image was reduced in size, we saved it as "smalltimbernerslee.jpg." Now we have two images: this one, and the larger image with which we started, "timbernerslee.jpg."

The following is the HTML code for a single thumbnail link, from the smaller version of the Tim Berners-Lee image to the large version that gives a more revealing view.

```
<a href="timbernerslee.jpg"><img src="smalltimbernerslee.jpg"
height="131px" width="200px" alt="Tim Berners-Lee thumbnail">
</a>
```

Here we have embedded an image source inside an image link, so that when we select it we go to the larger JPEG. As an exercise, resize an image you have by making it significantly smaller, then save it, naming the file so you know it is the thumbnail. Link it to the original or larger image using the code provided as a template, then see how it changes when selected in a browser. Make sure that the two image files and the HTML file are in the same folder.

Margins, borders, and padding

To understand how we can best lay out our content, we need to be able to apply the CSS and HTML tools so we can put padding, margins, and borders around our content. These features can be described as follows:

Content–The area that the text and graphics show up in.
Padding–The space around the content that separates it from the border. Padding takes on the background color of the content it surrounds.
Border–The area that visually distinguishes the content from both the padding and the margin. Unless the border is set to 0 pixels and thus does not show up, it is usually different in color from the padding and margin.
Margin–The space between the outer edge of the border and the rest of the content on the Web site. The margin is transparent and will be the color of the background of whatever surrounds the margin.

Review this graphic to familiarize yourself with how these features are defined:

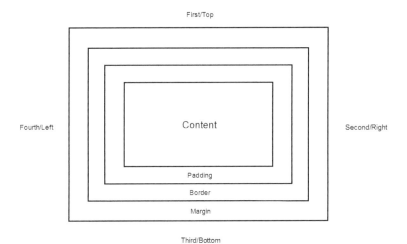

Figure 3.44 Content, padding, border, and margin.

In the examples that follow, we will work through the CSS and HTML code that describes these features. In Chapter 2 we briefly discussed how cell spacing, cell padding, and borders worked in a table and in table cells, or divisions, and these are similar concepts.

Margins

When setting margins for the <body> of a Web site, we can use simple HTML inline style coding, as we do here with the file from the section on images:

```
<body style="margin: 75px 30px 10px 30px">

<img align="right" src="purpleflower.jpg" alt="purple flower">

<h3>Everything You Need You Can Find In Your Neighborhood</h3>

<p>This is a flower in my back yard and it looks like it is getting ready to bloom. You never know what you will find close to where you live; you just have to open your eyes and look.</p>

<p>We found this flower in the morning before the dew had evaporated.</p>
```

The order of the margin coding in this example is as follows: the first value, 75 pixels, is for the top margin of the box; the second, 30 pixels, is for the right margin; the third is for the bottom margin; and the fourth for the left margin. This format is reflected in the box model diagram, where the first through fourth values are arranged in a clockwise order.

A more efficient way to control the margins around the content area of a Web site is to use CSS to set the body selector. Here we can explicitly name the margins, so the order is not important:

body {margin-top: 75px; margin-right: 30px; margin-bottom: 10px; margin-left: 30px;}

Alternatively, we could use the following CSS for the same effect:

body {margin: 75px 30px 10px 30px;}

Again, this simplified version of the margin coding follows the same format as the basic HTML coding in the previous inline style example. All three methods yield the same result:

Figure 3.45 Controlling body margins.

Compared to the file that we used in the images section, the margins in this file have been increased. The margin at the top is 75 pixels, which is why the image and the heading have moved down. The right margin is 30 pixels, the bottom margin 10 pixels, and the left margin 30 pixels.

Margins do not have to be restricted to <body> tags. We can use them in something as simple as a paragraph with a heading, as in this CSS:

p {margin: 10px 10px 10px 50px;}

h3 {margin: 25px 10px 10px 50px;}

Here is some HTML for a heading and paragraph (with text here elided) that the CSS can be applied to:

<h3>Information Literacy</h3>

<p>According to the Association of College and Research Libraries (ACRL), knowing how to select .... Acquiring these skills is not always easy, but it is immensely rewarding.</p>

When the HTML is saved with a reference to the CSS, and opened in a browser, we get the following screenshot:

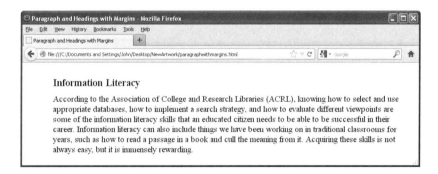

Figure 3.46 Controlling paragraph and heading margins.

The left margin for both the heading and the paragraph, <h3> and <p>, is 50 pixels and thus aligns both bodies of text. Review the other values to see how they result in the look we get in this screenshot. It is important to understand that margins can be used for different HTML tags, not just <body>.

Borders

Borders can be used effectively to draw attention to our text and images by distinguishing them from what surrounds them.

To add a solid border to the body tags we used previously, we would use the following code:

```
<body style="margin: 75px 30px 10px 30px; border-style: solid">
```

In the following screenshot we see a simple solid border:

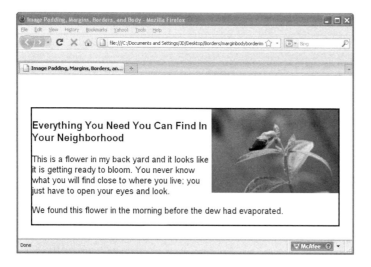

Figure 3.47 Controlling borders.

This code, "border-style: solid," provides us with a border that separates the text and image from the margin areas.

To better work with this file and the ones that follow, we will use a smaller image. Rather than "purpleflower.jpg," which you downloaded from this book's Web site, use "smallpurpleflower.jpg" and change the code in the "img align" line in the text box in the "Margins" section accordingly.

There are other properties and selectors for borders. For example, if we were to stay with the "border-style property," we could use the following values: "dashed," "dotted," "inset," "double," "ridge," and "outset." If we replace the value "solid" in the previous example with "dashed," we get the code "border-style: dashed." Using this would give us the following look:

Figure 3.48 Dashed border.

The "border-width" property allows us to engineer a border with a width appropriate for our needs. Adding the "border-width" property to expand the solid border to 5 pixels, we use:

<body style="margin: 75px 30px 10px 30px; border-style: solid; **border-width: 5px**">

This is how the change looks:

Figure 3.49 More emphatic borders.

The solid black border in this screenshot is wider and somewhat more emphatic.

The "border-color" property allows us to change the color of the border. We can use standard colors such as red, green, and blue, or use hexadecimal numbers to specify the color:

```
<body style="margin: 75px 30px 10px 30px; border-style: solid; border-color: blue">
```

Create an HTML file for the first example in this section and replace the <body> tag with this line of code. Verify in the browser of your choice that the border color is blue.

We can also put borders around other bodies of text; they need not be confined just to <body> tags, as we show in the following CSS code:

```
p {border-style: solid;}
h3 {border-style: double; border-width: 6px; border-color: blue;}
```

The screenshot shows how this code places borders around the heading and the paragraph:

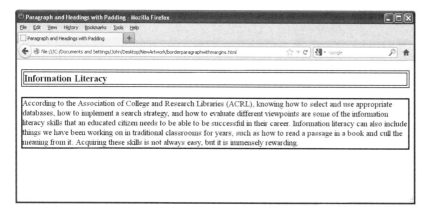

Figure 3.50 Borders around paragraph and heading.

Review these border properties and values and experiment with their effects using the body, h3, and p selectors.

Padding

Padding is the space between the border and the content. Study the following image source tags with inline-style padding:

```
<img align="right" src="purpleflower.jpg" alt="purple flower"
style="padding: 0px 0px 10px 10px">
```

Padding follows the same clockwise protocol as margins. In this example, we have added padding only to the bottom and the left side of the image; the top and right side are set to 0 pixels.

Figure 3.51 Padding around image.

Had we added padding around the image on the top and right side, the image would not be flush with the border; there would be white space between the image and the border at the top and right. Note that there is now more space between the image and the text than before.

In this code we are adding inline-style padding in the <h3> and <p> tags:

```
<body style="margin: 75px 30px 10px 30px; border-style: solid">

<img align="right" src="purpleflower.jpg" alt="purple flower"
style="padding: 0px 0px 5px 5px">

<h3 style="padding: 10px 10px 10px 10px">Everything You Need
You Can Find In Your Neighborhood</h3>
```

```
<p style="padding: 10px 10px 10px 10px">This is a flower in my
back yard and it looks like it is getting ready to bloom. You never
know what you will find close to where you live; you just have to
open your eyes and look.</p>

<p style="padding: 10px 10px 10px 10px">We found this flower
in the morning before the dew had evaporated.</p>

</body>
```

Verify in the code above that we have reduced the padding in the <img src> from 10 pixels to 5 pixels. The additive padding between the image and the two bodies of text will be 15 pixels.

Figure 3.52 Padding around image, paragraph, and heading.

This screenshot shows there is now more space between the border and the heading and paragraphs. Take some time to manipulate the padding, border, and margin values to see how it alters the layout and aesthetic of the content in this file. You might try to float the image to the left and recalibrate the values to make sure that it is appropriately aligned relative to the text.

In the following CSS code, we add some padding to the <h3> and <p> selectors:

p {border-style: solid; **padding: 15px 15px 15px 15px;**}
h3 {border-style: double; border-width: 6px; border-color: blue;
padding: 15px 15px 15px 15px;}

Combining this with the text we used previously, we get the following screenshot:

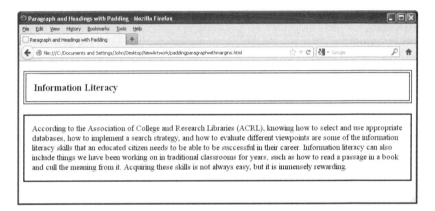

Figure 3.53 Controlling padding around paragraph and heading.

Note the space provided by the padding between the borders and the heading and paragraph.

Knowing how to use padding, border, and margin properties comes into play with many kinds of HTML and CSS elements. Go online and look at the CSS files associated with different Web sites (see the section earlier in this chapter on "Viewing CSS Files" for how to bring up the CSS code for a Web site). Review how many contain these properties, and how they are applied.

Additional CSS and HTML styling techniques

CSS and font properties

CSS allows us to select and control, and, if we are using external style sheets, to readily change the font family, style, weight, and size we use for our Web sites.

Font family

The "font-family" property indicates the particular font to be used. Times New Roman and Arial are two of the most common font families.

Times New Roman
Arial

The "generic font" describes the general appearance of font families, and browsers recognize five of these: "serif," "sans-serif," "monospace," "cursive," and "fantasy."

The two most commonly used generic fonts are serif and sans-serif. "Serifs" are the extensions that we find on the ends of letters. Times New Roman is a font family that has serifs, while Arial is a font family without serifs. That is, Arial is a sans-serif font, "sans" being a French word that means "without." The "R" in Times New Roman has serifs–there are two extensions, to the left and right, at the bottom of the vertical stroke that constitutes the left "leg" of this letter; one extension, to the right, on the right "leg"; and one extension in the upper left corner. The "R" in Arial does not have any extensions.

The typefaces in print texts have traditionally used serif fonts. Serifs help us discern letters more readily because "they tend to 'seat' the letters on the line and pull the eye along to the next word" (Schoff and Robinson 72). However, when it comes to headings in traditional print texts, sans-serif fonts have most commonly been used.

In online environments, where the resolution on a screen is not as acute as on paper, serifs cannot be as easily distinguished and it is generally thought that sans-serif fonts are more readable. Readability is about the medium.

In "monospace" fonts, each letter, number, or punctuation mark takes up the same amount of vertical space. Typewriters use monospace fonts in large part because if one needs to make a correction like replacing an "i" with an "o," the "o" will fit into the horizontal space originally taken up by the "i," even though the "i" itself is thinner. Computer code is also presented in monospace fonts such as Courier New.

"Fantasy" fonts such as Papyrus and **Impact** can be used to give an artistic effect, but are not as easy to read as other fonts. "Cursive" fonts such as **Comic Sans MS** and *Monotype Corsiva* are used to affect a handwriting style of text.

When coding our fonts in CSS, most often we use selectors such as <body>, <p>, or <h1> to <h6>, as these most readily demarcate sections of texts that will be assigned a specific text style. If you use <body> you are selecting a font style for all the text in your file. Here is how such coding looks:

```
h3 {font-family: Arial;}
p {font-family: "Times New Roman";}
```

For <p>, the selector is p, the property is font-family, and the value is Times New Roman. When we use the <p> tags in our documents with this

CSS code, the font will be Times New Roman. When there is a value that has more than one word in it, such as "Times New Roman," we put it in quotes.

Not all browsers display all of the font families that we specify, so it is best to list a series of font families in case our first choice does not work. When we are selecting these backup fonts, it is best to use the same generic family. In these examples, for <p>, we have three serifed fonts, and for <h3>, we have sans-serif fonts:

> h3 {font-family: Arial, Trebuchet, Verdana, sans-serif;}
> p {font-family: "Times New Roman", Georgia, Garamond, serif;}

If the browser someone uses to view our Web site does not have Arial, it will then go to Trebuchet for <h3> text. If not Trebuchet, it will go to Verdana. If none of these fonts are available, then some other sans-serif font will be employed by the browser, as we designate "sans-serif" as the last option. Here is some HTML to use with this CSS:

<h3>Font Families</h3>

<p>This is a sentence in a paragraph that is in Times New Roman, and the backup fonts are Trebuchet or Verdana. If these fonts are not chosen, then some other serifed font will be employed.</p>

Below is how the code will look:

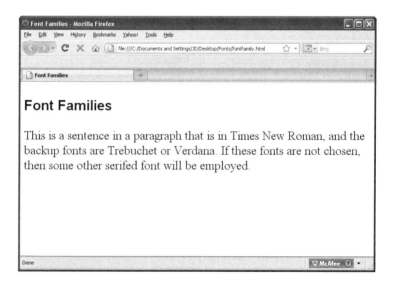

Figure 3.54 Font families.

Font style

The most common font styles that we can employ using CSS are "normal" and "italic." Most often italic is the one worked into CSS coding, as an alternative to the default normal.

Here is the CSS for making <h3> text italic:

h3 {font-style: italic;}

Note that "font-style" is the property, "italic" is the value, and "h3" is the selector.

We could also make "italic" an option when using a span in inline formatting (as we did in a previous section, "CSS Spans"):

.italic {font-style: italic;}

Using the "font-style" option in a span is a common way of marking italics. In HTML code without CSS we could use the <i></i> or <em></em> tags, which stand for "italics" and "emphasis."

Font weight

The most common use for the "font-weight" property is making an existing font bold.

h3 {font-weight: bold;}

In HTML without CSS, the <b></b> or "bold" tags and the <strong></strong> tags add weight to the text.

Font size

The "font-size" property determines, unsurprisingly, the size of the fonts we use. There are different ways to indicate the values for the font-size property. We can use pixels or px, em, points or pt, or percentages.

Table 3.1 Font size in pixels, ems, points, and percentages

Pixels or px	One pixel is one dot on your computer screen. We often describe screen resolution by numbers of pixels, for example "1920 by 1200" for a WUXGA screen. Like points, pixels give an absolute measurement, which does not scale up or down. Pixels are good for determining the absolute size of something on a screen. We used them when we determined the size of the header and footer sections, for example, in the "CSS and Layout" section of this book.
Em	The measurement "em" is equal to the parent font size, such as the font size specified in the CSS code for the "body" selector, so 2em is twice the default font size. If we have an 11-point font as default, a 2em size is 22 points. Ems are scalable, as they change relative to the given font size. As the screen changes in size, so fonts measured in ems will change in size, relative to each other. This "em" is not to be confused with the HTML markup tag <em>.
Points or pt	One point, or "pt," is $1/72$ of an inch. It is an absolute measurement, so text sized in points cannot be scaled up or down.
Percentages	When we use a percentage for determining font sizes, the current or default font is the 100% size—so if we specify 150% for another font, it will be one-and-a-half times larger. Using percentages means text is scalable for changing screen sizes.

If you were to designate that your font size should be 30 pixels in paragraph text, the CSS code would be as follows:

    p {font-size: 30px;}

When we are following best practice for producing a Web site, the sight-impaired should be able to readily see the text. Using ems for the font-size value works well here, as this means the text is scalable and thus meets the needs of a variety of screen sizes and users. If a user resizes the text on a screen and the fonts are in ems, all of the fonts–for example those used in headings and paragraphs–change relative to one another, and the viewer is able to see the text as intended, thus enhancing readability. If we specified absolute sizes using pixels or points, this would not be the case.

The minimum requirement, according to W3C's accessibility guidelines, is that text can be resized up to 200 percent without losing any information (W3C, "Resize Text"). This means that the text should be able to be made up to twice its size without overlap of content, and one word should not be so wide that it fails to fit on one line, "causing the sentence to be displayed as a vertical column of text that is difficult to read" (W3C, "Resize Text").

If we wanted to use ems to indicate the font size for the heading 4 and paragraph selectors, we would use the following, for example:

    h4 {font-size: 1.2em;}
    p {font-size: 1em;}

The heading 4 value is set to 1.2 times the paragraph value, which is 1em in height. Below is how text with this CSS coding looks in Mozilla Firefox:

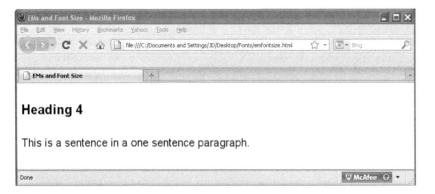

Figure 3.55 Font size indicated by ems.

In the following screenshot, the same Web site has been enlarged using the "Zoom" tool in Mozilla Firefox:

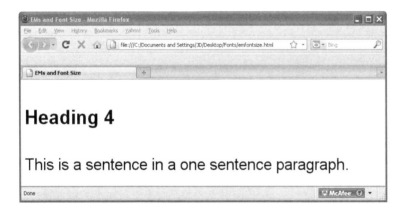

Figure 3.56 Font size using ems enlarged.

Both the heading and the paragraph texts have grown in relative proportion to each other and are thus more readable. This illustrates how ems can be used to produce scalable documents.

We could also use percentages in the CSS code to indicate the font size values:

h4 {font-size: 120%;}
p {font-size: 100%;}

Like ems, CSS percentage values are scalable.

Combining font properties

As we have shown in the examples of external style sheets, different font characteristics can be combined. For example, we could have the following for a paragraph:

> p {font-family: "Times New Roman", Georgia, Garamond, serif; font-size: 14 px;}

Here we are establishing both the font family and the font size. We can also combine these font properties with other properties, such as the line height and the font color:

> p {font-family: Verdana, Arial, Helvetica, sans-serif; font-size: 12 px; line-height: 1.5; color: #000000;}

Try out each of these CSS font selectors one at a time, applying them to a simple sentence using only the <p></p> tags in the HTML code, and see how they look.

Links and pseudo classes

Pseudo classes are another way of fine-tuning CSS selectors and classes. A pseudo-class element is connected to a selector with a colon (:) as follows:

> selector:pseudo class {property:value;}

Pseudo classes provide a different stylistic effect, often based on the dynamic state of the element. This is described in the section that follows.

Alternative link colors and styles

Browsers have default colors for links, but you can style them using CSS so they have different or "alternative" properties than the defaults. Color choice for links is important; if you choose a background color that is similar to the default link color, which is usually a shade of blue, viewers will not be able to see the link easily.

The pseudo-classes used for the alternative link color styles–"a:link," "a:visited," "a:hover," and "a:active"–are the most commonly used browser-safe pseudo-classes. Each pseudo-class is based on the dynamic state of the element; a different or dynamic look appears according to whether a user does not select, selects, hovers over, or visits a link. There are also other characteristics you can give to links using CSS in this way.

Here are the choices:

Link–This selector describes how the text you are using for a link will look if the viewer has not selected it yet.
Visited–This is the style the link changes to after it has been selected.
Hover–This is the style of the link when the viewer places the cursor over it but has yet to select it.
Active–The style of the link the moment the viewer selects or clicks it.

Here is some CSS code for each of these four selectors:

```
a:link {font-weight: bold; text-decoration: none; color: #6633ff;}

a:visited{font-weight:bold;text-decoration:none;color:#330099;}

a:hover {text-decoration: underline; color: #9685ba;}

a:active {font-weight: bold; text-decoration: none; color: #ff0000;}
```

The declaration "text-decoration: none" is used to instruct the browser not to underline the link's text, and this is one of the more common declarations on Web sites today; links look less distracting in this simpler form. The declaration "text-decoration: underline" is used to place a line under the link. In this case the text is underlined only when the viewer hovers or mouses over it; the unvisited and visited links will not be underlined. You can use other text-decoration values such as "overline," "line-through," and "blink." You might experiment with them, but these are rarely used.

Here is some simple code for links:

```
<body>

<p><a href="linksexercise1.html">Link</a></p>

<p><a href="linksexercise2.html">Visited Link</a></p>

<p><a href="linksexercise3.html">Hovered Over Link</a></p>

<p><a href="linksexercise4.html">Active Link</a></p>

</body>
```

Here are the links in a browser:

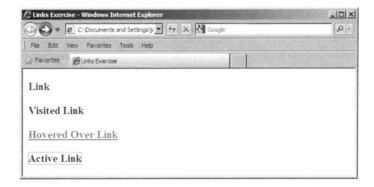

Figure 3.57 Links in dynamic states.

To become better practiced in this technique, choose a different background color and find some complementary link colors that are easy to see. Make sure that the chosen color for each of the dynamic states shows up well against the background.

Pseudo elements

Pseudo elements allow us to identify one part of some code and alter it for our stylistic needs. Pseudo elements are also pseudo classes.

First-letter and first-line

The first letter in every paragraph can be distinguished by using "first-letter" pseudo-class code, as in the following CSS example:

```
p:first-letter {color: #9933cc; font-size: 180%;}
```

This means that the first letter of every paragraph that is coded with <p></p> tags in an HTML document will have an altered color and it will be 180 percent the size of the other letters in the paragraph. Here is some HTML code:

<p>The first letter of every paragraph can be distinguished by using first-letter pseudo-class code as in the following CSS example: </p>

<blockquote>p:first-letter {color: #9933cc; font-size: 180%;}
</blockquote>

<p>This code dictates that the first letter of every paragraph that is encapsulated with paragraph tags will have an altered color and it will be 180 percent the size of the other letters in the paragraph.</p>

Below is how this code looks in a browser:

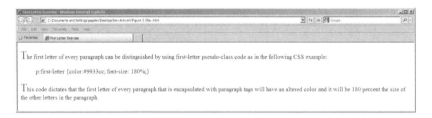

Figure 3.58 First letter pseudo element.

Note that the blockquote in the CSS code starts with a <blockquote> tag, not a <p> tag, and thus the first letter of the blockquote is the standard, default size and color. The two paragraphs, in contrast, start with a letter that is larger and a different color.

We can alter the first letter and turn it into a dropcap with the following CSS code:

p.dropcap:first-letter {font: 200% times; float: left; margin-right: 5px; margin-bottom: 1px;}

Compared with the previous first-letter code, we are using a different syntax:

selector.class:pseudo-class {property: value;}

Because we have inserted a class following the selector in this syntax—that is, "dropcap" after the "p"—we will need to alter the paragraph tag in the HTML code:

> <p class="dropcap">The first letter in this paragraph has a capital letter that is "dropped." It is 200 percent the size of the default font size for this paragraph and has a right margin of 5 pixels and a bottom margin of 1 pixel.</p>

Here is a screenshot of this dropcap example:

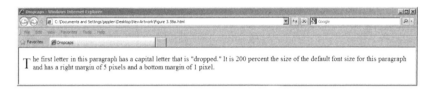

Figure 3.59 Dropcap pseudo element.

Contrast how this dropcap looks with the first letter in the previous screenshot.

The pseudo-element code "first-line" renders the first line of a paragraph in the particular styling you desire. The lines that follow below it will not have the same styling. This is much better than, in the following example, using HTML tags like <b></b> for bold around what we imagine will be the first line of a paragraph because the dimensions of browser windows vary screen to screen depending on the browser, browser settings, and screen size, thus the length or number of words in the first line will also vary. If you use the first-line pseudo-element code, the first line of any paragraph will always look the way you want it to regardless of how many words are in it.

Here is the CSS for the first-line selector and some specific styling:

    p:first-line {font-weight: bold;}

Here is an elided version of a paragraph we can use in the HTML code:

    <p>The pseudo-element code "first-line" renders the first line of a paragraph in the … regardless of how many words are in it.</p>

The <p></p> tags are all that is needed, as we did not designate a class in the CSS code. Here is one version of the file:

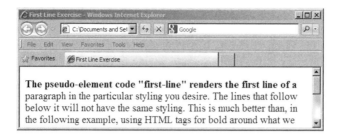

Figure 3.60 First-line pseudo element in small screen.

The first line is in bold and relatively short. Here is a second version in which the first line is longer:

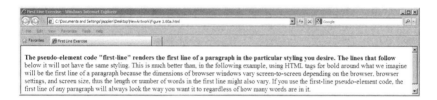

Figure 3.61 First-line pseudo element in larger screen.

Had we used the <b> tags in the HTML, we would not have the dynamic result that we get with the first-line pseudo–element code.

Color codes and backgrounds

There are several techniques used to indicate the colors we desire for backgrounds and other elements on our Web sites.

Hexadecimal coding is designated by a total of six letters, six numbers, or a combination of six letters and numbers, and it is preceded by a # sign. For example, white is #FFFFFF. Black is #000000 (with zeros, not the letter O). Many of the colors in between black and white are indicated with a combination of letters and numbers. For example, #E0E0E0 is the hexadecimal designation for a light gray. Hexadecimal or "hex" values instruct browsers to combine red, green, and blue in specific proportions to get the desired color.

We can also name the specific red, green, and blue combinations with a different syntax, one that uses "RGB" values or numbers. RGB stands for red, blue, and green. For example, the equivalent of #E0E0E0 in RGB values is "rgb (224, 224, 224)." Both of these give us a light gray color, and you can convert RGB to hexadecimal values or hexadecimal to RGB values using tools on various Web sites dedicated to color codes.

Sometimes using predefined colors is an easy way to indicate our choice, but the drawback is that there are only sixteen of them: black, white, maroon, yellow, blue, red, silver, gray, navy, purple, green, lime, teal, olive, cyan, and magenta.

There are many online sources for Web colors and their attendant hexadecimal or RGB values. When in doubt, the W3C Web site is a safe bet when you are deciding on colors.

Here is some CSS for the body and paragraph backgrounds:

    body {background-color: #E0E0E0;}
    p {background-color: white;}

The code for the body background color is a hex value and the code for the paragraph background is a predefined color, white. We could also have used RGB syntax:

    body {background-color: rgb (224, 224, 224);}
    p {background-color: white (255, 255, 255);}

Here is an HTML paragraph:

    <p>The background of this sentence is in white. It stands out against a gray background.</p>

Here we have a screenshot showing the gray <body> background with the paragraph on a white background:

Figure 3.62 Body and paragraph background colors.

We could also use simple HTML <body> tags, rather than CSS, to specify a background color:

    <body bgcolor="#E0E0E0">

Background images

Producing background images requires another style technique, and we start with a JPEG, GIF, or other image file. In the following CSS code, we are using a simple .jpg image. The formula for this code is "background-image: url(name_of_the_image.extension)":

> body {**background-image: url(purpleflower.jpg)**; margin: 150px 40px 10px 40px;}
> p {font-family:Verdana, Arial, Helvetica, sans-serif; font-size: 24px; line-height: 1.5; color: #000000;}

This is the HTML for a sentence that will show up as text in front of the image:

> <p>This is a flower in my backyard and it looks like it is getting ready to bloom. You never know what you will find close to where you live; you just have to open your eyes and look. </p>

Here is the screenshot showing this text and background image in a browser:

Figure 3.63 Background image with tiled effect.

Some backgrounds work well with the simple "background-image: url(name_of_the_image.extension)" code as we do not get the tiled effect. These backgrounds are based on images and are generally designed so that, when they repeat, the background looks homogeneous and seamless. However, the screenshot here displays a repeating background that does not reflect the best qualities of our chosen image. Additionally, the text does not read very well against this background. These two issues are addressed in the next section.

Making an image transparent

Images can be made transparent with one of the many photo-editing software packages available. Changing the opacity of an image can mean it works well as a background image, because the text stands out better and is more readable.

The following steps describe how Adobe™ Photoshop™ can be used to turn an image into a transparent background image (note that there are many other photo-editing packages you can use):

1. Pull up the photo-editing software and select the image file that you want to modify.
2. Double click on the "Background" layer in the lower right corner so that the "New Layer" dialogue box displays.
3. Move the opacity slider back from 100 or manually type in an opacity value other than 100. Adjust the percentage as needed, then select "OK."
4. Go to "File" in the upper left corner, select "Save as," and save the image using a new name that reflects that the opacity of the original has been altered. You do not want to replace your original image as you might need it later.

Below is a screenshot of the photo editing software with the necessary tools:

Figure 3.64 Adjusting image opacity values.

After the image with the new opacity value is saved, we need to add "no repeat," "background," and "fixed" instructions to the CSS code so we do not get the tiled effect:

```
<head>

<title>Background Image, No Repeat</title>
<meta http-equiv="Content-Type" content="text/html;
charset=iso-8859-1" />

<style type="text/css">
body {background-image:url(superlargepurpleflower.jpg);
background-attachment:fixed;
background-position:center;
background-repeat:no-repeat;}
p {font-family: Verdana, Arial, Helvetica, sans-serif; font-size:
24px; line-height: 1.5; color: # 000000;}
</style>

</head>

<body>

<p>This is a flower in my backyard and it looks like it is
getting ready to bloom. You never know what you will find
close to where you live; you just have to open your eyes and
look. </p>

</body>
```

You will probably have to resize your image so it will cover the browser window. To do this, you can employ the skills described in the earlier section "Thumbnail Image Links," but instead of diminishing the values for the height and width of your image, increase them. In the CSS code shown, we have done this with "purpleflower.jpg" and are now using the "superlargepurpleflower.jpg" file in the "background–image:url" property.

With these changes in the opacity of the image and the addition of the CSS code above, we get the following screenshot:

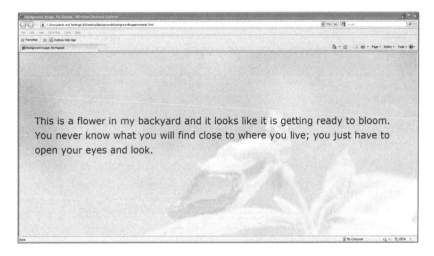

Figure 3.65 Reduced background image opacity and no repeat.

We now have a background image that stretches the length and width of the browser window and text that we can read.

Comments, metatags, special characters, and file transfer protocol

Comments

Sometimes you might want to put some commentary within the HTML coding of a Web site so others who view it can see why you did what you did, or when you did it. When Web sites have multiple architects, this can be helpful, as they can leave notes for each other about coding decisions. If you use the correct tags, the comments will not show up on browsers.

If you want to put comments between the HTML tags for your Web site, you need to use the following code:

    <!– –>

Here is an example:

    <!–Web site last updated on March 4, 2012.–>

For CSS coding, you need to use a different comment syntax:

```
/* */
```

The following is an example of this code with a comment on an ID:

```
#menu ul li {list-style: none;}
/* "#menu ul li" removes the bullets in an unnumbered list */
```

These kinds of comments have the added benefit of demonstrating that you are aware of a coding issue, and have thought it through and come up with the best way of solving the problem.

Metatags on web sites

Metatags are used between the <head> tags of Web sites to succinctly describe the site's key elements, and sometimes who has produced or authored the Web site. There are metatags for the description, keywords, and author of a Web site. Not only are they notes for people who like to view the coding of Web sites to see what they are really all about, they also help direct search engines to the Web site.

Here is an example from the *New York Times* of the code for a description, using the proper HTML syntax:

```
<meta name="description" content="Find breaking news, multi-media, reviews &amp; opinion on Washington, business, sports, movies, travel, books, jobs, education, real estate, cars &amp; more.">
```

In this code, we have the "keywords" metatags:

```
<meta name="keywords" content="United States Politics and Government, Israel, Iran, Obama, Barack, Murders and Attempted Murders, Bales, Karilyn, United States Defense and Military Forces, Iraq War (2003-11), Bales, Robert, Afghanistan War (2001-), Santorum, Rick, Romney, Mitt, Presidential Election of 2012, Paul, Ron, Gingrich, Newt, United States Politics and Government, Drug Cartels, Politics and Government, Mexico, Kidnapping, Movies, Social Networking (Internet), Productivity, Workplace Environment, Labor and Jobs, Organization, Flight
```

> Attendants, United Airlines, Travel and Vacations, Labor and Jobs, Airlines and Airplanes, Sculpture, Monuments and Memorials, South Dakota, Crazy Horse, Gowanus (NYC), Doomsday, Inventions and Patents, Madagascar Institute, Art, Hackett, Chris, Labor and Jobs, Syria, Middle East and North Africa Unrest (2010-), Movies, Goldman Sachs Group Inc, Slater, Steven, Smith, Greg, Suspensions, Dismissals and Resignations, O'Brien, Conan, Television, Labor and Jobs">

Note that the metatags for news organizations change quite a bit. One can imagine that some of the "keywords" here would not be in these metatags by the following year, as people would not be searching for them.

Many institutional Web sites do not include authors in their metatags. Because there are so many reporters and editors who work for the *New York Times*, it would be impractical.

Here are the metatags, including author metatags, for the Tim Berners-Lee Web site described in Chapter 6, as well as some comments between <!– –> tags:

<!–The image of Tim Berners-Lee is used with the permission of the W3C, and the W3C logo is used in accordance with this organization's permission practices. All other information copyright the author. The HTML/CSS coding and the writing and research have been done by the author.–>

<meta name="description" content="The early career of Tim Berners-Lee and the invention of the Internet, URIs, technologies of communication in place throughout history and how each of them transfers and restructures information and meaning, and the potential for invasion of privacy and abuse on the WWW."/>

<meta name="keywords" content="Tim Berners-Lee, Berners-Lee, Sir Tim Berners-Lee, URIs, URLs, inventor, invention, inventor of the Internet, Internet, SGML, HTML, URIs, protocols, HTTP, CERN, W3C, World Wide Web Consortium, privacy, abuse, open source"/>

<meta name="author" content="J.D. Applen"/>

Special characters

Sometimes we have to work a little harder when coding in HTML to produce what we call "special characters," which are symbols, diacritics, or punctuation marks. (Diacritics are those characters we add to a letter to indicate that it has a different sound, like the "~" or "tilde" we often see in Spanish words.) Sometimes when we use a special character such as an ampersand or "&" in our text we do not get the effect we want in all browsers. Thus it is best to get into the habit of using special characters if we want to produce browser-safe HTML code, as we do in the sentence below:

> <p>“The Great Atlantic & Pacific Tea Company—which is more commonly known as ‘A&P’—was the first grocery store chain in the United States,” explained Professor René Muñoz.</p>

The syntax for the special character code starts with an ampersand and ends with a semicolon, and between them there is some kind of abbreviation for the special character we are using. For example, "ñ" would be the code for a tilde over the letter "n." Here is how the sentence above would look in a browser:

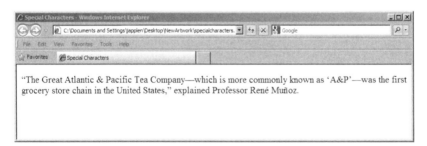

"The Great Atlantic & Pacific Tea Company—which is more commonly known as 'A&P'—was the first grocery store chain in the United States," explained Professor René Muñoz.

Figure 3.66 Special characters.

Usually, special characters such as em dashes and quotes show up in a browser, but characters like ampersands and tildes over letters do not, thus we should use this code. The following table shows a few of the special character codes, and there are many more:

Table 3.2 Special character codes

Name	Code	Character
ampersand	&	&
trademark symbol	™	™
left single quote	‘	'
right single quote	’	'
left double quote	“	"
right double quote	”	"
leftward pointing arrow	←	←
rightward pointing arrow	→	→
non-breaking space		
pound symbol	£	£
colon	:	:
semicolon	;	;
less than symbol	<	<
greater than symbol	>	>
en dash	–	–
em dash	—	—
copyright symbol	©	©
lowercase "n" tilde	ñ	ñ
lowercase "o" umlaut	ö	ö
uppercase "E" acute accent	É	É

For the tilde, we have to indicate in the code what letter we are placing it on top of. In the previous example, the tilde over the "n" in Muñoz is indicated by having the "n" follow the ampersand (&), and preceding "tilde;". Similarly, for the umlaut, we have an "o" following the ampersand and preceding "uml;", so the character will be an "o" with an umlaut above it.

The non-breaking space code is " " and this is an important special character for layout in HTML files, because it is a convenient way to make a space and it is browser-safe. In HTML, just hitting the "space" key does not work, so we need to use a " " special character code for this effect. (Note that in the "Character" column in the table the non-breaking space does not show up, as its purpose is to produce a space.)

File transfer protocol (FTP)

To move copies of your HTML/CSS files onto the Web for everyone else to see you are going to need, first, some space for your files on a server, and, second, an FTP client, which is a software package that will transfer files from your hard drive to the server. Once on the server, your files will be accessible to everyone who has your Web site's URL.

The screenshot shows an FTP client for Mozilla Firefox, FireFTP, which was constructed by Mime Čuvalo. You can download it at http://fireftp.mozdev.org.

Once you have this, go to the opening screen of FireFTP and choose "Edit":

Figure 3.67 FireFTP "Edit" button.

Type the requisite information into the dialogue box. In this example, the "Host," or server, is "pegasus.cc.ucf.edu." The "Account Name" can be whatever you want—perhaps your name and a shorthand note that reminds you what server your files are on would work, as shown:

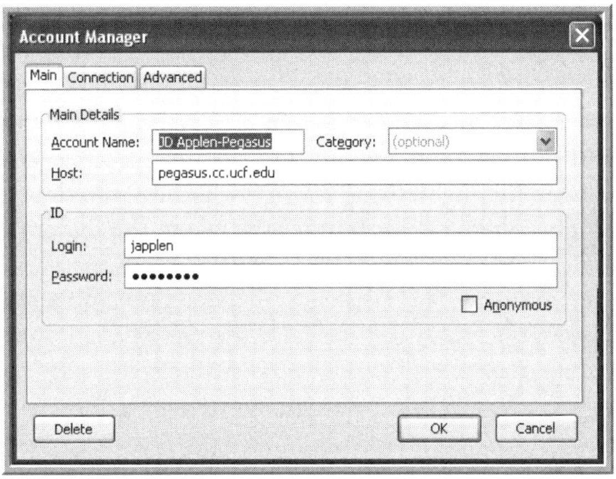

Figure 3.68 FireFTP "Account Manager" dialogue box.

You will also need to type a login name and password into the relevant fields. Putting the host name and the login together, the URL for this particular site is pegasus.cc.ucf.edu/~japplen. This is what you would type into a browser to view the site. Click "OK."

On the full FireFTP screen, select "Connect":

Figure 3.69 FireFTP "Connect" button.

You will come to a screen that looks like this:

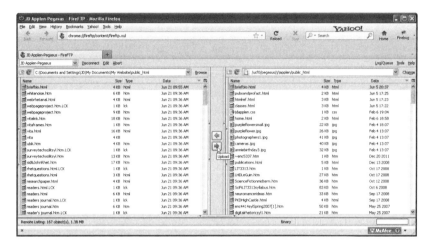

Figure 3.70 Transferring files from hard drive to server.

On the left side of the screen shown in the screenshot are the files on the hard drive, and on the right side are the files on the server. The "briefbio. html" file has been highlighted and the arrow pointing to the server side has been selected. This will move or transfer this file to the server from the hard drive.

While there are many FTP clients that you can use, most of them follow this method for transferring files and have the same basic layout.

Works cited

Berners-Lee, Tim, and Mark Fischetti. *Weaving the Web: The Original Design and Ultimate Destiny of the World Wide Web by Its Inventor.* San Francisco, CA: Harper Collins, 1999. Print.

Manovich, Lev. *The Language of New Media.* Cambridge, MA: MIT Press, 2002. Print.

New York Times. "Metatags." *New York Times.* Web. 18 March 2012.

Schoff, Gretchen and Patricia Robinson. *Writing and Designing Manuals: Operator Manuals, Service Manuals, Manuals for International Markets.* Boca Raton, FL: CRC, 1991. Print.

W3C. "Resize Text: Understanding SC 1.4.4." Web. 2 June 2012.

Chapter 4

Rhetoric and writing

Chapter overview

The content of Web sites is largely based on the written word, and an awareness of rhetoric allows us to tailor our words to meet the needs of our audiences. In this chapter we will learn to apply rhetoric to our writing, identify writing styles most appropriate for the Web, and discuss the use of other resources and how to document them to better present our ideas.

Rhetoric

Classically, rhetoric has been known and defined as "the art of persuasion." This stems from the practice of rhetoric in the political arenas of ancient Greece and Rome, and the major figures in early rhetoric were Isocrates, Socrates, Plato, Aristotle, and Cicero. Any time we are using texts to explain something in the world in some way, we are adding some kind of interpretation of what we are trying to explain. Beyond the choice of words, organization, and style, rhetoric also involves the implementation of templates of logic (*logos*) that have been tested and seem to be "reasonable," the establishment of authority in the speaker or writer (*ethos*), and the use of information or ideas that naturally stir the emotions (*pathos*).

For our purposes, we need to think of ourselves as rhetoricians, or *rhetors*, when we write text for, choose images for, and organize our Web sites. While we might not be standing and giving speeches before others in ancient Greece, we still need to think about the way we present our information to be competent Web site architects.

How we use rhetoric

We are always acting rhetorically, as we are always adjusting our discourse or language for different audiences. When we first meet someone, we are usually a bit more formal in our behavior and language; perhaps we do this to avoid causing offence, thus giving ourselves some time to take the measure of our new acquaintance. In time we get to know the person and we usually loosen up a bit and venture a joke or even tease our newfound friend.

Rhetoric takes many forms other than words. How we choose to look can be understood as a kind of symbolic or rhetorical discourse. There is a subset of young adults that have entertained the practice of body piercing and experimenting with less than conservative hairstyles. This is perhaps a rhetoric of disaffection that represents a challenge to the controlling nature of parents, teachers, government officials, and any other site of power in our society. It also gives us a feeling for the ethos of the person. What we wear has much to do with the kind of authority we project and this also changes with the audiences we are confronting. Some people respond positively to the look just described. Some people do not respond well to it, and would prefer to believe someone who looks a bit more conservative, perhaps a man in a business suit or a woman in a pant suit.

Rhetorical moves can also be made at the level of diction, or word choice. For example, businesspeople who are trying to win approval for new housing refer to themselves as "developers," and to the structures they want to build as "homes" in a "community." This makes them look like they care about people and their very human need for shelter. When they call themselves "developers," they are suggesting that they are taking something wild and imperfect and making it better for humans. Those who oppose the developers, their environmentalist detractors, will refer to the same structures as "units" that will exist in a "tract" and will displace a "natural habitat" or "ecosystem." They are trying to point out that there is something very impersonal about the physical structures that the developers want to put in place of something that is beautiful and irreplaceable, has been there for a long time, and should not be destroyed.

Ethos

Rhetoric is used to persuade people, to show them that the rhetor—the speaker or writer, or for our purposes the Web site architect—is right. In many instances, rhetoric is used to demonstrate the quality of the ethics of one's position, or the ethos of the rhetor.

We can also see the way these practices work in the media and in public events. For example, what is the rhetoric of a news anchor who never takes her eyes off of us as she peers into the camera and relates the day's events? Some have suggested that the "eye contact" of a news anchor makes us feel that she is talking to us in a very personal way, and thus she can be believed more readily; the anchor's ethos is enhanced. Why is it that, for a time at least, there were young reporters on the NBC national news, usually in shirtsleeves, scurrying around in the background of a set or looking into computer monitors as the anchor read from a teleprompter? How does this enhance the authority of the news anchor? Does it suggest that those people in the background are working hard to find the most recent news, the "newest" news, and that if a story breaks during the broadcast, it will be passed on to us through the anchorperson in the course of the half-hour show? Perhaps it suggests that there is more than one person, the anchor, out there finding out about the news; instead, there is a huge organization that is both literally and figuratively "behind" the anchor and supporting him or her. Could it also suggest in a rhetorical fashion that what the anchor is saying is based on sound sources? Are there other ways that this can be done? How would this practice be translated for the needs of a writer? For our purposes, how does this rhetorical ploy from television news translate to online news Web sites? Can you find some analogies?

The importance of rhetoric in technical communication and Web site design is something we all need to think about. When we produce Web sites, we cannot count on people seeing what we are wearing or hearing our mellifluous voices as we speak, and they will not see the expressions of concern or warmth on our faces. We can theorize, however, about how effective choice of words and sound logic better allow us to produce a successful message because they have enhanced our ethos; people respect and believe us.

The rhetoric of the phrase "Mistakes were made" allows bureaucrats to subtly hide from their mistakes by transferring fault to someone else in their organization. What they are really suggesting is that there are two ideas at play: "I am in charge and take full responsibility, but someone else created this mess." These arguments undercut each other, but this rhetorical ploy is still commonly used.

As Web site architects, we need a working knowledge of rhetoric because it allows us to answer the following questions:

- How do we come across as a people? What is the ethos we project when we are producing online texts built with our words and images?
- Can we tweak our message to alter our authority or ethos and thus enhance the efficacy of our message?

Knowing the ethos we project allows us to better connect with our audience.

Situated ethos and invented ethos

There are two kinds of ethical proofs or constructs of ethos: situated and invented.

A situated ethos is when the speaker, writer, or Web site that represents an organization already has a reputation, preferably a good one, one that can be relied on by the audience. This significantly enhances the prospect that whatever is in the speech or text or on the Web site will be thought of in a positive light and believed, or at least considered. In the age of Rome, coming from "a good family" had much to do with how your character was perceived. In fact, it was important to be from a good family to be a member of the ruling class. The family name gave rise to situated ethos; someone from a good family would be thought of as someone people could trust. This is something still seen today: having Adams, Bush, Clinton, Kennedy, or Roosevelt as a last name does not hurt one's chances of being elected to national office.

Over time, people, corporations, and institutions have established themselves as trustworthy, and thus it could be said that their latest efforts can be welcomed or trusted. People have a faith in Jimmy Carter and Nelson Mandela, as they have established a long track record of commitment to good causes. Some people would not think of buying anything but an Apple computer, and others a PC with a Microsoft operating system. The Humane Society of the United States and the United Way each have a situated ethos that many admire and support without question. However, the situated ethos of these people and entities has been established over time and is the product of an originally invented ethos.

An invented ethos is one that the rhetor has to establish for her audience when they first meet her, hear her speak, read what she has written, or come to her Web site. This is the ethos that most of us have when we produce Web sites, as very few people are so famous or have such a reputation that everyone already has an opinion about them. This is also true of companies you might be working to represent in your role as a professional communicator.

There are three ways that we can enhance our own ethos or judge the ethos of someone else. First, we need to believe that the person has practical experience on the matter in question so she will make the right decision or present an issue knowledgeably; competence must be earned and projected. Second, the person must not have a character flaw that can be discerned, because we cannot be sure that the position she is presenting is not one based on self-interest; she must be of "good character." Third, we need to know that a person will present the best case for us, and not one that she knows is inferior. She must have "goodwill" (Crowley and Hawhee 201).

One of the ways we can believe in someone's ability to make informed choices and present an issue competently is by knowing that the rhetor has had experience with an issue. People who have been confronted with a problem and overcome it, or who at least better understand the problem because they have been confronted with it first hand, are more likely to be believed. Parents who testify before government officials about their inability to find adequate and affordable medical care for one of their children would be an example of people who have a personal basis for their knowledge. Credentials are another way of establishing credibility. Professionals with advanced academic credentials and publications or who can claim expertise by having worked in a field for a significant period of time are more likely to be believed because they have shown they have studied the issue at hand. Often an appropriate use of some jargon used to characterize the nature of an issue is helpful, but if someone goes overboard and uses language in such a way that it makes absolutely no sense to the audience, he runs the risk of not being accorded the ethos he is striving to invent.

According to Cicero in his *On the Parts*, good character is based on "virtues," most notably "generosity, sense of duty, justice and good faith" (quoted in Crowley and Hawhee 207). While the first-century Romans were different from us in many ways, the value of these character traits still holds up today; the work of public relations professionals so often creates a vision of celebrities, politicians, and even corporations by celebrating these same traits. To be generous, to devote your time to others and charity, is a common marker of someone or some institution with good character. Celebrities in all arenas often say that they are performing charitable works as they want to give something back to their community or country. Politicians and military professionals will characterize their efforts as based on a sense of duty to their country. Justice and fairness are also elements of the persona of people in the public eye and serves as a basis for all the decisions they make. If someone seems to understand basic fairness, all of their actions can be thought of as those of someone with good character, as their good deeds would serve to help people regardless of their backgrounds; they are fair minded. Even if they might have made a mistake, their intentions were honorable, as they acted in good faith.

In *On Invention*, Cicero adds "the ability to bear suffering" as a virtue that enhances good character (Crowley and Hawhee 207). Public figures have also related past experiences about periods of loss and difficulties in their lives to show that they have suffered, have become stronger, but have not forgotten what it is to have been in difficult life circumstances. Some politicians like to position themselves as members of society who have had to work for everything they have, or who perhaps lost a child or family member to accident or disease.

The modern criteria for securing goodwill can be identified in presentations where the rhetor presents his or her ideas along with just enough information to meet the needs of the audience without offering superfluous or unneeded detail. Crowley and Hawhee point out that audiences have faith in movie reviewers who present just enough information about a movie to allow them to figure out whether or not they should pay to see it, without telling them the ending and thus spoiling it. Additionally, goodwill can be established when a speaker or a writer describes just why what the reader will be reading will be of benefit to them—authors oftentimes do this in the preface to a book—or what the benefit of accepting a certain argument or position will be (Crowley and Hawhee 210–212).

Pathos

Pathos, or a "pathetic" argument, is when we insert texts in communication which arouse the emotions of the people we are trying to persuade. This is evident in the use by non-profits of pictures of animals that have not been treated well or children who are starving in foreign countries. Some say that pathetic arguments are weak because they involve the emotions and not reason, but others would say that when we use only cold, hard reason, we forget what is really important. Sometimes it is a good idea to stir the emotions of our audience about things that really matter; today we are bombarded with so much information, and much of our entertainment industry is built around comedy that, while clever, is built on a cynical worldview in which everything is funny and nothing really matters. However, if we use pathos in a way that obstructs the issue and keeps people from also using reason, or logos, and from considering the authority, or ethos, of the rhetor, the use of pathetic examples is irresponsible.

Pathos is best understood as dealing with emotions such as anger/calm, love/hate, fear/confidence, shame/shamelessness, envy/emulation, and compassion, joy, and hope. Pathos works best when it is put into play in a manner that gets people to feel that their values are being undermined or that someone close to them is being threatened in some way. Pathos is not about what Cicero would refer to as "appetites" such as pleasure and pain (Crowley and Hawhee 246–247).

In ancient times, pathetic proofs were thought of as a responsible way of persuading people of a belief or idea; in contemporary society we are often told "That's just an emotional argument" which should not be considered. Logos, or logic and reason, is considered a more appropriate use of rhetoric today. However, the ancients often looked to pathos as a heuristic, that is, a method you use to figure something out, which allowed them to better examine the logos and to define an argument. For example, Gorgias believed that pathos could be used to put people in the appropriate state of mind so they could more clearly understand certain arguments. Perhaps this is truer today than some admit (Crowley and Hawhee 251). For example, for us to understand the "reasons" for contributing money to aid people in, say, Darfur or Somalia, we first need to see pictures of children who are refugees or who are starving. After this, we are more receptive to the "logic" or empirical evidence; given a deeply felt sense of the suffering of a handful of children in the pictures we see, the numbers that reveal that many thousands of people are suffering have a real effect and are not just abstractions.

We can also see pathos at work in support of ethos. When giving the State of the Union message, it has become fashionable for American presidents to look to the gallery of citizens seated in the House of Representatives chamber and to point to and publicly commend the act of some brave soldier who has risked his life in military action or some courageous citizen who has rescued someone from drowning in freezing waters. How does this enhance the authority of the president? Some would say that in this case a president is using pathos to enhance his or her own authority or ethos. We respond to heroes in an emotional way, and if a president points out something like this, we might admire her for appreciating the same kinds of ethical behavior that we appreciate. In this case, pathos is used to enhance ethos. One might ask, is this responsible? The presidents who do this are not the heroes when they do this; the heroes are the heroes and the presidents who use this technique are hoping that some of the good feeling these people inspire will rub off on them. However, perhaps it is appropriate that our presidents remind us from time to time that there are real heroes in our society, and perhaps this rightly enhances a leader's ethos, as she or he knows what is important.

To use pathos effectively, a rhetor first needs to know the present state of mind of the audience members, and she needs to know what will move them. Following this, the rhetor needs to understand the reasons why the audience will be moved (Crowley and Hawhee 251).

We can illustrate how to employ pathos via this three-point rubric in an extended example applying the pathetic figure *enargeia*. Enargeia involves narrative or description of an event so vivid that the audience members feel it is taking place before their eyes (Crowley and Hawhee 258). Michael Pollan (47) describes a process by which the overuse of fertilizer is destroying the environment using this figure; not only does he describe the scientific effect, we can see it:

> The ultimate fate of the nitrates that George Naylor spreads on his cornfield in Iowa is to flow down the Mississippi into the Gulf of Mexico, where their deadly fertility poisons the marine ecosystem. The nitrogen tide stimulates the wild growth of algae, and the algae smother the fish, creating a "hypoxic," or dead, zone as big as the state of New Jersey—and still growing. By fertilizing the world, we shrink the planet's composition of species and shrink its biodiversity.

Not only does Pollan describe the scientific effect, he describes a movement of a "nitrogen tide" that ends up destroying an area as large as a well-known Eastern state with a substantial industrial base. He could have made it more "pathetic" by focusing on a single fish that had washed up on the shore of the Gulf of Mexico, but perhaps he decided that his audience would not need this image as the phrase "smother the fish" is enough. Where the present state of mind of the reader is one in which fertilizer is thought of as something that nurtures and does not destroy, using words such as "dead" and "hypoxic," and the phrase "deadly fertility poisons" challenges this belief. Overall, the audience can be moved by and clearly "see" a large deadly body of poison moving from an idyllic Iowa cornfield down the Mississippi river, a national treasure, and, ironically, not nurturing life but destroying it.

Irony is itself a powerful rhetorical device; if something is revealed to us in an ironic way, the opposite of what we imagined to be true is the truth, and this shakes us up or challenges us to reexamine an idea, event, or situation. In this example, the irony is that something seemingly so natural and nurturing is actually killing living things.

Logos

Logos is the Greek term for logic, and it is thought of as a way of explaining our ideas to others using reasoning. Also, it is thought that by applying logical forms we can uncover great truths. The classic logical form is the syllogism:

> Major premise: Socrates is a man.
> Minor premise: All men are mortal.
> Conclusion: Socrates is mortal.

It would be nice if we could find a way to apply tidy syllogisms like this to everything that troubles us in the world today, but we cannot. All arguments have logics, and some are easier to discern than others. The following are some logical arguments from contemporary culture and politics and the premises that support them.

On U.S. involvement in the Middle East:

- **Argument 1** If we don't go over to the Middle East and vanquish terrorism, terrorism will come here to America.
  Premise Because of the 9/11 attack, we know that there are people in the world who will travel to America and attack us, and there have been subsequent attempts to attack us on American soil. This dramatic event in our history still lives with us, so this enforces this argument. What if we go where the perpetrators of this event live, train, and plot attacks against our country and bottle them up there? We will be safer.
- **Argument 2** By keeping our armies in the Middle East, we will continue to stir up so much anger that people who were never our enemies before will become our enemies and fight us until we leave.
  Premise While not downplaying the gravity of the 9/11 attack, we have to understand that if an invading army came to the United States and stayed for over a decade, American citizens would band

together to throw out the occupiers. This is what has happened in the Middle East and we are now seen as occupiers. It is best if we just defend our own shores.

On embryonic stem cell research:

- **Argument 1** You cannot destroy life to save it.
  Premise Embryonic stem cells could eventually become a human being, thus to use them to help fully developed human beings who may be ill is not ethical. What is key here is the potential and the "right" of these cells to develop.
- **Argument 2** An eight-cell embryo that stem cells are taken from does not constitute a life, and we can save real lives with this research.
  Premise These cells may have begun the process of development toward becoming a fetus, but they do not yet constitute a human life. We could use cells at this stage to cure fully developed human beings who are ill.

The best way to understand logic is to understand the basis for the major and minor premises it is built on. There are as many logics as there are issues, and intelligent and honest people can find different logics to support what they believe in, based on the assumptions they make. For example, one group of people will base their logic on the "right" of a set of cells to develop, while another group will base theirs on the idea that cells at this point are "not human."

Organization

Organization is a tool of rhetoric, and there are many organizational strategies; just two basic organization strategies are described here.

In the first, you can present all of the facts of a situation and then present the thesis. If you have laid out the facts in the right order—A → B → C—they only can lead to one conclusion, your thesis. It will seem like you have analyzed the situation correctly and you are right.

The second basic organizational strategy starts with an idea or a thesis, as in the classic essay form, then the writer illustrates in point-by-point fashion how the thesis is supported by the facts.

Both methods have their strengths and weaknesses and both adopt hierarchical or linear forms of organization, which we describe later in this book.

When we are given an essay, we think we need to read it sequentially to read it correctly. If we are presented with a written biography, it is implied just by the form that we need to read it sequentially to get a complete understanding of who the biography is about.

In a hypertext structure, there is not the linear approach that we find in the two classical organizational styles, and thus Web sites encourage only partial readings. Because of the variety of patterns, there might be a different rhetorical effect for each reader.

As is described in sections that follow, we layer information on Web sites into pathways that allow for a more user-friendly experience. This is somewhat like a table of contents in a book, but a layered organization scheme allows us to believe that we do not always have to carefully read every point in a hypertext document to find the information we need.

Kairos

Kairos is one of the most complex and all-encompassing concepts in rhetoric in that it demands that we consider all aspects of a situation as we analyze it. Perhaps the most appropriate phrase in English that captures the concept of kairos is "rhetorical situation." A rhetorical situation is one where an issue or idea is understood in its historical context and in the context of the power relationships between the speaker or writer and their audience.

There are two important elements to kairos, the temporal and spatial. The temporal element is the timing of a message; a good rhetorician has to have a calculated sense for when his audience is ready to hear something. Those in favor of more regulations for banks and other lending institutions knew the timing was right for financial reform after the steep drop in stock market equities and housing prices, with so many voters losing their savings, homes, or jobs. So politicians advocating reform would be reasonably safe in doing so—the voters were ready to hear this; the timing was right.

Where something is said or done can be considered the spatial element, but "spatial" can also be understood in other ways. Who says it and when it is said is important, but the positioning of the rhetor between different camps and the forum in which something is said or written, can also be considered aspects of the spatial element.

One can clearly see the elements of kairos in what is now referred to as Barack Obama's "Race Speech" of March 18, 2008. This speech was given by President Obama to clarify his position on race in the wake of some public comments by Reverend Jeremiah Wright, the African-American minister of Obama's church in Chicago. Because Wright made some comments that challenged his perceptions of the political power of white America, Obama had to distance himself from Wright, but do so respectfully, as he did not want to lose the support of African-Americans.

Obama initially situates himself as a man who has ancestors and family members who are both black and white:

> I am the son of a black man from Kenya and a white woman from Kansas. I was raised with the help of a white grandfather who survived a Depression

to serve in Patton's Army during World War II and a white grandmother who worked on a bomber assembly line at Fort Leavenworth while he was overseas. ... I am married to a black American who carries within her the blood of slaves and slaveowners—an inheritance we pass on to our two precious daughters.

Not only is he a man of mixed descent, as so many Americans are, his ancestry is deeply embedded in American history. They are patriots as they served their country during World War II. They are also people who have suffered through the Depression and America's cruelest institution, slavery. As he positions his ethos, Obama better places himself to make his points. His father is identified as someone who was not born in the United States, and most Americans can go back and find an ancestor who also immigrated to this country.

Obama also seeks to place the incendiary issue that started with Wright's comments in the context of the history of American politics, which has seen divisive issues set Americans against themselves and kept them from moving forward.

For we have a choice in this country. We can accept a politics that breeds division, and conflict, and cynicism. We can tackle race only as spectacle—as we did in the OJ trial—or in the wake of tragedy, as we did in the aftermath of Katrina—or as fodder for the nightly news. We can play Reverend Wright's sermons on every channel, every day and talk about them from now until the election, and make the only question in this campaign whether or not the American people think that I somehow believe or sympathize with his most offensive words. We can pounce on some gaffe by a Hillary supporter as evidence that she's playing the race card, or we can speculate on whether white men will all flock to John McCain in the general election regardless of his policies.

We can do that.

But if we do, I can tell you that in the next election, we'll be talking about some other distraction. And then another one. And then another one. And nothing will change.

While this can be understood as a temporal element of kairos, it is also spatial, as Obama is showing how many elements from different political positions come together in the American media landscape; they are from people who range from liberal to conservative, and people who are white or black or something in between. To be clear, this is different than the cliché that so many American politicians use, the cliché that "We are all Americans and we need to pull together and support our country." Instead, Obama positions himself so he can be seen as recognizing the differences and that we are not all the same, but we need to move forward in spite of Wright's remarks; at this moment, he argues, if we do not understand this event for what it is and the many different positions that it stems from, we cannot go forward and the issue of race will drag us down. The way he positions himself in the previous passage about his parents and heritage helps him to establish an ethos that allows him to deliver this argument convincingly.

What is most important here is that kairos underscores our need to understand that reality is socially constructed when it comes to understanding ideas and contemporary events. All of the elements of a rhetorical effort—ethos, pathos, and logos—come together in a certain temporal and spatial arrangement, and this is what gets people listening to it and acting on it—or, if the rhetor's use of these elements and timing is off, ignoring it.

Figurative use of language

Perhaps the most interesting and powerful form of persuasion can be seen in the metaphorical use of language. For instance, the line "the voice of your eyes is deeper than all roses" (cummings 44) does not make much sense on a literal level, but it can read as a very powerful suggestion on a figurative level. It is from a poem by e.e. cummings, a poet who was averse to using capital letters, which is perhaps another rhetorical ploy. Because only a few people explain the power of the figurative in concrete fashion, an extended example follows.

Perhaps "the voice in your eyes is deeper than all roses" seems like something someone might say to a partner who he or she is close to, maybe at the end of the day, perhaps an hour from sundown. There is something beautiful, serene, and melancholic about this time of day that makes it easier for people to be together. The idea of this coming about in the late afternoon starts with the "deeper than all roses" part of the line. This phrase suggests the image of the shadows that show up *deep* down inside of a rose between the petals, where the color gradually darkens. This, along with the "eyes," might lead one to imagine the shadows that show up on our faces during the twilight hour; if we don't run to turn the lights on when things start to dim, if we just allow the gradual darkening to slip into our lives, it seems that the contours of our features stand out a little more. There is not that washed out look we have at noon when we are walking to wherever we are going to eat lunch.

The eyes figure heavily because this is where the shadows seem to have their most dramatic effect. At a figurative level, perhaps it is possible to hear the "voice of [someone's] eyes." When the line is read aloud, there is a simple elegance in the way it sounds. There are ten words in, and all save two are monosyllabic, and those two, "deeper" and "roses," each have only two syllables. These are simple words, words that people use when they are relaxed and honest, and all of them together suggest the soft-spoken integrity of people who respect one another. When read aloud, they seem to gather a gentle momentum that makes them all the more meaningful. The "s" sounds in "voice," "eyes," and "roses" work to suggest an equation between them. And if you look at the way these three words are spaced throughout the line and the way the "s" sound announces them, you get an interesting rhythm that perhaps points out their importance ("the *voice* of your *eyes* is deeper than all *roses*").

The meaning of the words in this line is extremely important, too. "Voice" and "eyes" tell us that something very human is involved here. And in our culture the word "roses" is most often associated with love and passion. What really amplifies the role of roses in this sentence is the word "all." This line tells us that the *depth* of "all" the roses in the universe does not measure up to the profound feeling, the *depth*, of the quiet nobility and humanity of the eyes we sometimes experience in our lives.

Of course, this is just one way of explaining this line, but it demonstrates that in a very restricted textual space—ten words—a lot of rhetorical power is sometimes unleashed. However, there are symbolic, figurative texts that we do not examine with care—we just imagine that they are saying something important to us and we do not need to challenge them. For example, the narrator in Kurt Vonnegut's *Breakfast of Champions* challenges us to think about "the picture of the truncated pyramid with the radiant eye on top of it" that is on our dollar bills. Of this iconic image in American culture, which everyone has seen but few have really thought about, the narrator tells us, "It was as though the country were saying to its citizens, 'In nonsense is strength'" (9). There are some texts in our culture that need to be examined as they sound good or look authoritative, but perhaps are based on "nonsense."

We all have our own favorite lines that, when examined, are based on whimsy and emotion, but they stay with us because they mean something in a different way than cold logic can play out for us. Imagine a line from a song that you have heard or a snippet of a conversation you had with a friend that you remember because it has a power in it that separates it from all of the other useless bits of information that swirl around us in our lives. There are figurative texts that we are affected by that have a real rhetorical power and sometimes they should be examined because we accept them without question.

The following is the opening line on the National Park Service Web site for Yosemite National Park: "Not just a great valley, but a shrine to human foresight, the strength of granite, the power of glaciers, the persistence of life, and the tranquility of the High Sierra." What does this mean? Why is it a "shrine" and how does this word suggest that the park is anything different than a group of rocks, trees, cliffs, waterfalls, and streams? What did "human foresight" have to do with the coming into being of a place in east-central California that had been in existence for many thousands of years before humans came along and gave it a name? Words like "strength," "power," "persistence," and "tranquility" also suggest religious discourse. Unpack this tidy figurative description and, after viewing the graphics and some of the written information on this Web site, how does it serve to distinguish Yosemite from other places on Earth?

Exercises

1. What is the ethos of the *New York Times* Web site (www.nytimes.com)? When you first see it on your computer screen, does it convey an ethos that would make you feel that the "news" it presents is accurate, important, and up to date?
2. The *New York Times* (NYT) at one time was referred to as "The Grey Lady" because the earliest versions of it did not have etchings or photographs: it was six columns of written text. To some, this actually enhanced the ethos of this paper, as it suggested that it was not about fluff but rather information that could be depended on, an "all business" ethos. Contrast the copy of the NYT at www.nytstore.com/ Original-New-York-Times-From-First-Year-in-Existence_p_6565. html with the online version of it today. How is it different in terms of ethos? What does the NYT offer now that it did not when it was referred to as "The Grey Lady"? Then contrast the online version of the NYT to online versions of other newspapers, such as the Los Angeles Times (www.latimes.com) or your local paper. What kind of ethos is projected?
3. Go to the Web site of a major corporation, your university, a major public figure in entertainment or politics, a non-profit organization, or to the personal Web site of someone you know. How do these Web sites project the ethos of the organization or person? Can you break down the ethos into its situated and invented components? How might they be improved?
4. What kinds of figurative texts exist in your local community that you have not examined, but have some power? Why do they have the power they do? What are the rhetorical elements involved?
5. When do we use emotional arguments or pathos in our writing or choice of graphics? What kinds of examples do we provide to make points that might be thought of as elements of pathos? In the example about heroes from this section, pathos stirred the emotion of admiration. But pathos can also stir other emotions, such as fear and anger. When is it appropriate or ethical to use pathos?
6. Find the Web site of a major car manufacturer and describe how ideas about the environment have been employed to sell cars. How does the manufacturer seize the opportunity to be environmentally friendly? What elements of kairos are in play here?
7. Identify the major elements of rhetoric—ethos, pathos, and logos—in the quotes from Barack Obama provided in this section. How do these major elements contribute to kairos?

8. Describe the computer "desktop" metaphor, a kind of figurative language, and how it has been appropriated by computer manufacturers. Is it really like a desk you might sit at? What about the "Trash" or "Recycle" icons on the desktop that better allow us to understand where it is we throw things out? What other icons contribute to the building of this metaphor?
9. Go to the Appendix and review the Sullivan Ballou letter. What components of ethos are in play? Identify the religious elements. Do they elevate Ballou's prose? Detail the features of pathos. How might pathos make Ballou's letter an effective text for understanding the suffering endured by people in war? How is pathos used here to move us in a way that logos could not?

Audience

"Know your audience" is a commonplace or classic refrain that composition instructors have been using for ages, and all the texts we produce should follow this general rule, whether they are essays for an English or history professor or the text you place on your own Web site. "Audience" has been defined in many ways, and typically we look at demographics such as gender, age, income, nationality, ethnicity, and education level when we try to understand who our audience members are. Depending on the substance of what you are trying to communicate, one might ask questions that determine where people live, their political party, profession, and home ownership status. Other questions could include how many hours they spend watching television per week or how much time they spend online.

Crowley and Hawhee provide additional insights into the makeup of audiences and how they are likely to receive messages. These insights are based on the work of classical rhetoricians and some contemporary studies of audiences.

Regarding technical issues, general audiences not only need to be informed about the matter at hand, but the ethos of the writer or speaker needs to be such that the audience is convinced that she knows the issue (Crowley and Hawhee 202).

Rhetoricians also have to tie what they are speaking or writing about to the needs of the audience, or, at the very least, make it interesting to them. Just stating the points of a situation and explaining it, no matter how correctly, will not have an effect on an audience if its members do not find it important to them in some way (Crowley and Hawhee 300). More than this, rhetoricians can work to make the audience realize the benefits of listening or reading, so they are more likely to listen (Crowley and Hawhee 212).

General audiences hold one of three attitudes toward the rhetor: hostile, indifferent, or accepting. The likelihood that an audience member will change their mind is based on their "emotional intensity" regarding the issue.

Additionally, the way someone identifies with an issue is also an indicator of how they will hold on to an opinion (Crowley and Hawhee 254). For example, if someone considers himself to be "an environmentalist," his position on global warming or offshore oil drilling will be firmer than someone whose sense of self does not include being an environmentalist.

In rhetorical studies, the concept of a "discourse community" allows us to ask questions about audiences in specific ways. A discourse community is a group of people who write, listen, speak, and think through an issue using a certain set of discursive practices. "Discourse" can mean any "sign" that communicates, including graphics, music that accompanies a video on a Web site, or the way you might dress if you were being interviewed for a job. Most commonly, discourse means the words and sentences we speak, read, write, and hear.

A discourse community could be a group, such as technical communication students at a large American university, members of the medical community, or members of the U.S. Navy who work on aircraft carriers. Within these discourse communities there are usually other subgroups. In the Navy example, there are officers and enlisted men, pilots and airedales—the enlisted men who help launch and service the aircraft. In the medical community, there can be nurses, surgeons, and pharmacists, among many other groups.

When we write for a discourse community we have to be aware of the language that its members will understand, and this is a classic way of determining who your audience is. Of course, if we use jargon, we need to make sure it is jargon that our readers will understand. But a discourse community takes this idea farther, and to understand this we need to go into some more theory.

Language is "intertextual," which means that it is built on traces of previous languages. In our own lives, we first use the language that we learn from listening to our parents, but as we spend more time with our peers we alter the way we use discourse and begin employing the words and phrases of our generation. The lyrics we so often hear in rock music can be traced back to the lyrics used by Mississippi delta blues artists such as Robert Johnson. The same can be said of many of the musical structures that rock musicians employ. These could include the chord progressions and rhythms of Robert Johnson that eventually ended up influencing the work of artists such as Eric Clapton. James Porter has shown that much of the discourse or language used in the Declaration of Independence, while literally a revolutionary document, was based on the discourse of political philosophers whose writings preceded it. For example, the phrases "Life, Liberty, and the pursuit of Happiness" and "all men are created equal" existed and were in use before they flowed out of the tip of Jefferson's pen. Discourse conventions are built on the traces of previous discourses, and all discourse is intertextual (Porter 39).

Based on the idea of discourse communities, James Porter extends the notion of audience. Because language is intertextual, we can ask more penetrating questions regarding discourse. For example, in this view, the audience also shapes the structure of the discourse. In the past, the writer was thought to consider the audience, then write for its members; there was no awareness of how the conventions of the audience shaped the writer. It was believed that the writer was in complete control of the discourse and that language is "autonomous"; the language emanated from the writer and no one else (Porter 41). It was a one-way transaction, from writer to audience. A better understanding of this transaction is that the writer and the audience interact with each other: "If we regard each written product as a stage of a larger process—the dialectic process within a discourse community—then the individual writer's work is a part of a web, part of a community search for truth and meaning" (Porter 43). While Porter is not necessarily using the word "web" in the same way that is used in "World Wide Web," which is the focus of this book, we can extend his thinking to writing for the Web. We have additional influences on us as writers based on our interactions with the assumed audiences that will be reading our online work.

Knowing this allows a writer not only to understand but also to write within a discourse community. A writer's words are shaped by the discourse community. For example, Jefferson might have thought that he would have to use phrases like "all men are created equal" because the people he wanted to convince that it was time to break away from Great Britain had also used and grown comfortable with these phrases and would not challenge them. Eric Clapton might have taken a blues chord progression that had been around for some time and changed it for his own needs with the idea that people would have some degree of familiarity with these already tested and elemental musical structures. In using these discursive conventions for their own needs, Jefferson and Clapton both shape the thinking of and are shaped by the people for whom they were producing texts. The point is that writers need to know the audience's conventions, but they are also shaped by these conventions as they write or compose within a discourse community.

Instead of just collecting "demographic data" like age, political party, profession, or gender and using these data to write for an audience as the all-knowing writer, we should also ask questions that fall within four general areas: Who speaks/writes? To whom do they speak/write? What do they speak/write about? How do they say/write it? (Porter 46). In asking these questions, we get closer to the heart of the matter. Not only do we understand the audience, we understand how we both shape and are shaped by the audience, and the traces of language that compose us.

Who speaks/writes?

According to Porter, we need to ask "Who is granted status as speaker/ writer? Who decides who speaks/writes in the forum? By what criteria are speakers/writers selected?" (46). With this approach we are in many ways returning to the classical sense of ethos and how it is established, and there are also elements of power involved. If someone is writing for the *Journal of the American Medical Association*, only in rare cases will the author be someone other than a medical doctor, the primary criterion for being selected. This is pretty obvious.

When a technical communicator creates a description of a process that will appear on her company's Web site, she is given some leeway in the way she emphasizes the ideas because she is the chosen writer. Her company thus selects her to represent the company, but then her work has to be overseen by others who will read her drafts, so her writing in some ways is negotiated with others. This negotiation is reflected in the now acceptable use of the word "we" by technical communicators who write for large organizations; this will be discussed in a later section. The writer also has to use other words and phrases that are in play in the discourse community within which she is writing. For example, the phrase "going forward" is used in the financial industry to mean that what has happened in the past is something we cannot do anything about, in fact we might have even done something wrong, but now we are doing the best we can by giving you this instruction and advice: "Going forward, you should have a more conservative portfolio."

Who the "important figures" in the forum are, and what the "key works" or sources are that need to be cited, are also important elements in distinguishing a discourse community. Each field in academia and medicine has its own key figures and works that people often feel they always have to draw attention to in their work to get published, regardless of their real value. There are stock athletic and political heroes that people in the public eye also refer to in an effort to better gain an audience's support, and American politicians will often cite the Bill of Rights or a political document for the same purpose. These phenomena can tell us a lot about the discourse community.

To whom do they speak/write?

When Thomas Jefferson was selected to draft the Declaration of Independence by a committee that included Jefferson himself, John Adams, and Benjamin Franklin, this too was an interactive negotiation, and Jefferson knew he was writing for others who could be rather critical: the committee that chose him, the audience of American citizens, and, more widely, people in France who the colonists were looking to seek aid from if they were to go to war with England. This was a complex discourse community. Existing discourse within this community was fair game here and could be used by Jefferson in writing for this audience, but the audience and the existing discourse were also having an effect on Jefferson.

Porter identifies forums of assembly as an element in a discourse community, and with this in mind we might imagine that the people we are writing for get most of their information from professional organizations or online chat groups populated by people in the field. In the workplace, they might work problems out with their immediate colleagues in face-to-face environments, rather than through email, Skype, or blogs. Outside the work-place, perhaps some people get most of their ideas from their family members or from the religious institutions to which they belong. These are all forums, and they have different assumptions and rules. Most importantly, within these forums there is a give and take and communication does not go one way. The members of each community engage in discourse, built on traces of language, which is accepted and shaped by all the other participants.

Additionally, Porter (46) points to attitude, prejudices, background, and level of proficiency, experience, and knowledge with respect to subject matter as key factors in discourse community analysis; these are also the kinds of criteria classical rhetoricians would use to identify audiences.

What do they speak/write about?

Regarding "valid proofs" and "methodological assumptions," academics and professionals in the medical field, for example, often do not use online wikis as their primary source of information and will only consider something written in an academic journal as worthy of their trust because the manu-scripts that end up in these journals have been analyzed by outside reviewers who are experts in the field. A pharmacist might believe that a drug is only effective if it can be shown through double-blind clinical studies with a large pool of participants that it actually works. Others will believe in the curative power of a medicinal product if an actor goes on television and tells them that it worked for him. Some go to wikis because just about everyone else is doing so, and anyone can contribute to them; there is an implicit sense of openness and democracy in wiki environments, and this helps ensure that the information presented is accurate.

What are allowable subjects in a forum is also something that is important to consider. There was a time when substance abuse was thought to be a character defect best treated by a psychologist and that people who failed in overcoming it were weak willed. However, medical research has since revealed that people can inherit this habit; for many it is genetically pre-determined. Thus substance abuse can be written about in forums where it is understood as an inherited disease and that unfortunate people who succumb to it should be treated as patients, not moral inferiors.

How do they say/write it?

Language should be thought of less as an expression of each individual's unique worldview and more as a means by which we relate to one another. As we use language to express ourselves to others, those others need to recognize the words, phrases, and expressions that we use. If they do not, we will know it quickly, as they will not give any indication that they understand. Language that works is "socially constructed," meaning that the language we use today is successful only because people have a shared understanding that the words used within their discourse community actually mean something. You might recall having said something in a way that just did not work, then revising what you said through the diction, or word choice, or the structure of the sentences, so that it did work, meaning someone did or said something that indicated to you that they finally understood. In the example on substance abuse, the word "patient" casts a different light on people than the word "alcoholic," which for many has a derisive tone to it.

Kenneth Gergen (157) more completely defines this phenomenon and includes other symbols in the languages we use:

> Any action, from the utterance of a single syllable to the movement of an index finger, becomes language when others grant it significance in a pattern of interchange, and even the most elegant prose can be reduced to nonsense if others do not grant it the right of meaning.

Using your finger to make a "point" only works because in the past, when we have used our finger to make a point, people have understood what we meant. To put it another way, if we used our index finger to make a point and no one understood what we meant by it, we would find another gesture or word that our audience understood. Using phrases like "all men are created equal" had currency in Jefferson's day because others in Jefferson's discourse community understood what it meant. What gave this phrase its special effect when Jefferson used it was that he was not just writing an essay on political philosophy; he was saying that he and others were willing to go to war over what they perceived to be a "long train of abuses," or injustices.

The four questions discussed here are in many ways based on the work of the French philosopher Michel Foucault, who would describe these criteria as the basis of "discursive formations," the term he uses for discourse community in his book *The Archaeology of Knowledge*. He would go so far as to say that we are constituted by the intertextual statements of our culture that enter into our thinking, and which sit or tumble around in there in a manner that we are not aware of, then are reconstituted in our thoughts, actions, or speech. While this might sound a bit humbling, the healthiest way to think of this is to work hard to examine just how the discursive formations we are writing within work. In this way, we can write within them, but also know that we are writing within them and remain cognizant of the limits of these conventions. It also tells us something about who our audience members are.

Exercises

1. Imagine one specific discourse community that you belong to. It could consist of people in your major, people who engage in the same hobby you engage in, people who are from your home town, or people who have the same political inclinations you have or who belong to a religious organization you belong to. Go back through the questions described by Porter—Who speaks/writes? To whom do they speak/write? What do they speak/write about? How do they say/write it?—and identify some answers for this discourse community. Then identify some of the specific, concrete language and terms used in this discourse community.
2. After you have finished identifying the concrete terms used within a discourse community you belong to, think about how you would explain them to someone who is not in that community. How does this cause you to reflect on language you perhaps have taken for granted?
3. Find a Web site that you often visit that addresses a particular discourse community. Go back through Questions 1 and 2 and answer them for this Web site.

Audience and writing for the web

The preceding sections present the basics of rhetoric in order that the theories and theorists discussed in this section may be better understood. Understanding our audience or "discourse community" is the fundamental practice of rhetoric, as it allows us to better shape our communication with that audience.

Ginny Redish has provided us with some thoughtful techniques for understanding Web site construction and audience. Rarely do people come to Web sites for long reading sessions; they come to Web sites for information. They are looking for accurate and up-to-date information, information that allows them to complete a task, or information that they can readily find (Redish 3). To provide such information, we need to extend our understanding of ethos in oral and print literacy contexts and apply it to Web sites so we can convey to our audience the feeling that in our roles as Web site writers and architects we are presenting accurate information.

We have to know that our Web site visitors already have a pre-formed idea of what they need; they are looking for something that will allow them to finish something they already know something about. Because of this, they will often skim the information (Redish 2). However, in a contrast with Redish's emphasis, sometimes users read with great care. When people search a Web site, they muddle through to get to where they need to in as quick a fashion as they can, but when they land on the body of text or graphics that meets their informational needs they are likely to read it with some care.

Still, when we construct our Web sites it is prudent to assume that patrons of our work are only there to find the information they need quickly and then move on. For this vast majority, if we have not organized our information in a way that means people can readily identify what they need, we have failed. Jakob Nielsen points out that all too often people build Web sites that reflect the internal structure of an organization, not the users' needs: "Users should not have to care how your company is organized." Sites should be "user-centered" and "determined by the tasks users want to perform on the site" (Nielsen, *Designing* 198). We will discuss navigation later, but it is important to emphasize that this also contributes to knowing our audiences.

Redish asserts that strong Web writing can be characterized as a "focused conversation," where people with specific and perhaps even urgent needs come online with a problem or concern that can be posed in the form of a question (Redish 4). The concept of discourse community that we have been discussing allows us to imagine a "focused conversation," as we already have a sense for the language and the traces that build this community. We also now have an additional rubric that provides us with a sense of who will be listened to, who we can ask questions of, and the range of questions that will be thought of as relevant in this community.

If you think about it, we often turn on our computers because we have a question that needs answering. The first thing we usually do is go to a search engine or go to a Web site that we already know about that has been produced by someone we have some confidence in. If we work to square this with Redish's characterization, there is always a concern or feeling, perhaps only an inchoate or partially-formed feeling, that when further examined can be reduced to a question. When we go online and read a set of the FAQs, or "frequently asked questions," that we see on so many Web sites, this shows that others have understood the role of questions in Web site construction for some time. But strong Web site writing, according to Redish, anticipates the questions of the audience members who will be having a conversation with the site's content. For example, she characterizes several basic question structures, such as "How do I …?" or "Where do I find out about …?" (5).

When you go to a Web site, say for a university library, what would be some of the basic questions you might ask? "Where do I find out about information help desk hours?" or "How do I access online journals so I do not have to drive to school to do research?" As writers for a university library, we might have produced some content that addresses these concerns because we imagined having a conversation with a potential library patron.

To better understand what questions might be asked, we need to present the criteria that define the audience. While we have defined audience in classical rhetoric, we need to extend it to relate specifically to Web sites. Redish starts with a basic version of audience when she asks us to imagine what she calls "personas"—imaginary constructs of individual people who exemplify the general target audience members (20). A persona is a "some-one" you can imagine, someone who is supposedly less of an abstraction than a list of characteristics that make up a typical member of your audience. Jennifer Fleming has also used this technique, though she uses the word "profile" in place of "persona" (8).

Below is a recasting of the items Redish (20) lists to describe the information we should use to produce a persona:

- Demographics such as age, expertise, experience, and ability
- Technology
- Values
- Emotions
- Cultural and social environments
- Phrases and quotes that can be attributed to the persona.

Demographics might include age, gender, and what someone does for a living (Redish 18–19). For our purposes, technology could be the kinds of Internet access and screen sizes that an audience member uses (Redish 17). With respect to cultural and social environments, we might imagine that as they view your Web site people might be, for example, on a break at work without much time to read, or alone and not surrounded by others (Redish 18).

Redish takes the term "values" to mean "knowing what matters" to Web site patrons. For example, someone might be looking at your Web site because he has some important decisions to make and really needs some information. Someone else might need to know the cost of shipping a product before she considers buying it; while a customer service professional might be frustrated as he searches for information to help him ensure the Web site of the company he represents states its policy clearly so as to not confuse its patrons (Redish 17). Redish lists other emotions patrons might experience, such as "intrigued," "curious," "impatient," "skeptical," and "stressed" (16). While this is a limited list, it is important to be able to imagine other emotions and combinations of emotions that people have when they are online.

Thinking about some of the items that make up a persona, a female university student majoring in psychology who has done volunteer work with troubled teens for three years and who wants to be a child psychologist could be the basis of a Redish-style persona. Using this method, we would give the student a name, Jennifer Jones for example, and have someone pose for a picture that people designing a Web site could see so they could get a better sense of a real person with concerns and personal aspirations.

Adding to this, we would get a sense of the typical Web tasks that Ms. Jones would engage in, such as checking email, using a social media Web site, checking up on online courses, and reading her university's online version of the school newspaper; and we might imagine she does this all on a two-year-old laptop. She might want to be able to do her online course at one in the morning, as she gets off work late, and she wants to know both that everything on the course Web site is written clearly and that the technology can be counted on to be working, because she will not have anyone to call and ask questions. Additional information for her persona might include hobbies, part-time work, and that she engages in on-campus activities such as being in the Psi Chi International Honor Society; also that she is a psychology major because she ardently believes she can help people during difficult times, and the phrase "I think you will be able to work through this" captures something about her feelings toward people and what she is learning.

All of these elements go into the construction of an imaginary persona and help us get a sense for what the real audience might be for a Web site. Still, to be clear, this is an amalgam of traits that are distilled into one imaginary human who is supposed to represent the "typical" audience member. The problem, of course, is that no single person can represent an entire audience, though thinking of a person to represent the central traits of this audience can be helpful. For many, however, it might be better to think about who is in a discourse community and to ask the questions that James Porter asks in his approach.

Exercise

Describe in specific detail a Redish-style persona that you would use to think about before making specific content decisions for a Web site for your university that informs and addresses the needs of prospective students for your major, whether it is technical communication, information technology, or any other major. What are the specific hopes of this persona? What might he or she be concerned about? Make sure you include ideas based on Redish's approach, as described in this section. Refrain from reducing your answer to things like how many units a student has to take; get to the heart of the personal needs of the student.

Writing and style

Choosing the words we use is the basis of ethos, pathos, and logos—the fundamentals of rhetoric. When we generate written texts for our Web sites, the style of writing is equally important. Beyond presenting ideas with clarity, we need to be mindful of the rhetorical dimension of what we write.

Active and passive voice

We can write with greater clarity using the active voice because it readily identifies the "agent" in a sentence. The agent is the person or thing in the sentence that is doing something, and the "doing something" part of the sentence is described by an active and specific verb. If a writer leaves the agent out of a sentence or buries it deep within a sentence, meaning we have to think for a moment about who the agent is, we generally have a weaker sentence. The following are variations on active and passive voice constructions.

> **Active:** The students presented their ideas with vigor and clarity.
> **Passive:** The ideas were presented with vigor and clarity.
> **Passive:** The ideas were presented with vigor and clarity by the students.

The active construction in the first sentence celebrates the achievements of the students better than the passive constructions in the next two. Even though we are told in the second passive construction sentence that the students were doing great work, it takes its time to do it and buries the emphasis on the students.

> **Active:** We made some mistakes.
> **Passive:** Some mistakes were made.

In the passive construction, the agent(s) who made the mistake are not identified. In this classic example, the writer or the speaker could conceivably have used the passive voice to avoid identifying who made a mistake, thus avoiding responsibility.

Being direct means it takes less time for a reader to think about what you said.

At times, it is a good idea to use the passive voice to break up the syntax patterns in a paragraph, though generally it is a good habit to try and keep just about all of your sentences in the active voice.

However, the passive voice works well in some genres of writing such as report writing in the sciences.

> **Active:** We trapped, tagged, and released thirty-five bar-tailed godwits during this period.
> **Passive:** Thirty-five bar-tailed godwits were trapped, tagged, and released during this period.

Science writers usually keep the agent out of a sentence because this suggests a rhetorical objectivity. It does not matter who did the work; what matters is the accuracy of the data.

Variations of the verb "to be" often get in the way of more powerful verbs that we can use to produce stronger sentences. This is not to say that there is not a place for "to be" forms. The existential form of "to be," the verb form that we use to characterize our existence, possesses a power of its own.

When Hamlet asks, "To be, or not to be," he deftly questions the meaning of our lives.

"To be" variations include, but are certainly not limited to, the following words and phrases: are, is, has been, can be, will be, was, were, and should be. There are many more. There follow some versions of active sentences and their weaker passive constructions.

> **Active:** We need to build sentences that substitute stronger verbs and we can entertain Hamlet's questions later.
>
> **Passive:** Sentences can be built with the substitution of stronger verbs and Hamlet's existential questions can be entertained later.

The actors in the first sentence, "We," have been pushed out of the second sentence with a version of the "to be" form, "can be." "We" shows up in the first sentence and takes charge.

> **Active:** The faculty members support the students in all that they do.
>
> **Passive:** The students are supported by the faculty members in all that they do.

The "to be" form in the second sentence here is "are supported," replacing the stronger verb form, "support," in the first sentence, which is simple and direct.

> **Active:** The administrators have determined that the student population will increase by six percent in the next four years.
>
> **Passive:** It has been determined that the student population will increase by six percent in the next four years.

As in the first example in this section, the "to be" or passive sentence has undermined the actions of the actors, here "the administrators." In the active sentence, the administrators are making decisions and in command. The passive sentence puts more emphasis on "the student population," which "will increase," and this hides the actors, the people who are actually doing something.

Combining sentences

Some technical communicators believe that we should strive to write sentences that are short because they are easier for people to read. Each small sentence might by itself be easier to read, but a passage of small sentences can be rather difficult to read. Strive to combine most of your sentences. People often think through or talk through ideas using rather long, meandering strings of words that form their thoughts. If we reduce these thoughts into short, start-and-stop sentences where we have to continually keep identifying the subject of the sentence and what act it is performing it is like driving over a series of syntactical speed bumps.

Separate Sentences:
It is important to learn how to use commas, dashes, semicolons, colons, coordinating conjunctions, and conjunctive adverbs.
They are explained in the punctuation section of this book.
Not only do they allow us to combine sentences, they allow us to combine sentences so that they are easier to understand.

Combined Sentence:
It is important to learn how to use the commas, dashes, semicolons, colons, coordinating conjunctions, and conjunctive adverbs that are explained in the punctuation section of this book because not only do they allow us to combine sentences, they allow us to combine sentences so that they are easier to understand.

It is clear in this combined sentence which tools allow us to combine sentences, where they are explained, and what they do.
In this section we also discuss why writing wordy sentences can make it harder for your readers to understand your ideas. This does not undercut the value of combined sentences as long as we make sure that we cut out the unneeded words in them.

Using "You" and "We"

Redish pushes the use of "you" when addressing your audience members, even for "serious and important" messages, as it makes writing "inviting and personal" (172–173). For example, the Mayo Clinic professionals use "you" when addressing illness in a section entitled "Hospitalization":

If you and your doctor decide that you need hospital care, you will be admitted to Mayo Clinic hospital on campus. If you are too ill to be examined as an outpatient at the clinic, you may be admitted directly to the hospital.

Imagine this passage without the use of "you":

If a patient and the patient's doctor decide that he or she needs hospital care, the patient will be admitted to Mayo Clinic Hospital on campus. If the patient is too ill to be examined as an outpatient at the clinic, he or she may be admitted directly to the hospital.

Here the reader has to imagine himself as "the patient," which is a step removed from "you." If "you" happen to be too ill to be examined, it sounds like someone is speaking directly to "you" in "your" time of need. Being referred to as "the patient" is a little less inviting and more of an abstraction. The phrase "a patient and the patient's doctor" sets up a distance between the two people involved in the decision; it removes the "your doctor" from the sentence, a phrase that sounds comforting because it more strongly affirms that we have someone on our side we have chosen and can trust.

The use of "you" also allows us to compose sentences that do not have the "he or she" style of gender-specific phrasing (Redish 175). In the example, sooner or later we have to switch from "the patient" to "he or she" to keep from sounding repetitive.

Using words like "we," "us," and "our" to represent an organization is also an effective stylistic device because it allows people to feel they are in a "conversation" with the organization when they are reading ideas online (Redish 178–179). For example, we have the following from the Mayo Clinic to describe its newsletter subscription service and an email address requirement:

> We require you to provide an email address to subscribe to a newsletter. If you subscribe, we may also use this email address to send you communications about other Mayo Clinic products and services. We may also share this email address with our third party email management vendor for purposes of delivering the newsletter to you; however, we do not share your email address with groups outside of Mayo Clinic for any marketing purpose.

If we substitute the phrase "the Mayo Clinic" for every "we" that shows up in this message, the organization would have sounded impersonal. The use of "we" in this legal text allows the reader to hear the voice of people being methodical but human.

Use concrete language

Abstract language is flaccid and does not help us to convey concrete information. It might sound pleasing and easy on the ear, but it does not instruct. Here are some concrete examples of what this means. In the 2012 campaign for the presidency, Mitt Romney's campaign slogan was "Believe in America," and Barack Obama's was "Forward." Does "Believe in America" mean that Obama did not believe in Americans and what they can accomplish? Does "Forward" suggest that we would go backward if Romney was elected, or that if Obama was re-elected we would continue to march forward out of the recession? These slogans fail to tell us at a concrete level what the specific tax rates would be for a typical family, or how each candidate would determine who is and who is not an illegal immigrant. The phrases are also interchangeable. Romney could tell us that we need to move "forward," and Obama could say that he "believes" in Americans.

At the most rudimentary level, going from the abstract to the concrete requires that in each step you are further specifying a smaller and more unique subset of elements. For example:

vehicle → car → sedan → midsize → 2003 → Honda → Accord → four cylinder → standard transmission

There are many kinds of vehicles: cars, trains, bikes, motorcycles, ships, and planes. A car is a subset of vehicle, and sedans are a subset of cars. This does not mean you always have to say "Where did we park our 2003 Honda Accord midsize four-cylinder sedan with a standard transmission?" You can say, "Where did we park our car?"

When presenting information, it is usually best to choose a more concrete term or phrase over an abstract one. A phrase such as "achieving excellence as a student" is a useless cliché. Here is a more concrete explanation of what this means: "Students should go to all of their classes and labs and listen carefully to their instructors; read and understand the material as required in the syllabi before classes or lab periods; and go over difficult concepts until they finally understand what they mean and can apply them, or, if they cannot, meet an instructor in her office hours before or after class and ask about these concepts." If you do this, you are "an excellent student." This is why concrete description is helpful.

If you present an abstraction, follow it up with a more concrete example. Chapter 1 of this book provides a quote from Sven Birkerts in which he contrasts a word carved in stone—which "has weight, grandeur—it vies with time"—with a "weightless" word on a computer screen (Birkerts 155). This quote is followed up with an explanation: "If a stone carver took so much time to carefully chisel out some words in stone, we feel that these words are more important than words that have been dashed off and instantly show up on a computer screen because the stone carver puts in so much focused physical and mental effort." Birkerts has wonderful ideas, but sometimes some extra explanation is needed when characterizing them.

Wordiness

When we write a sentence that at first sounds great, sometimes it is not the best sentence. A good practice is to go back over passages you have written and see how sentences in them can be condensed. This is important because shorter sentences that say the same thing as longer sentences usually convey their ideas with greater clarity and rhetorical power.

Review this first version of a sentence from the Preface and compare it to the later, condensed, more succinct version:

> To be able to understand the "social, cultural, political, and educational values," I feel that we need to examine just how communication technologies present information and how they shape the information we find when using them.

Understanding the "social, cultural, political, and educational values" requires that we examine how communication technologies present and shape information.

The second sentence is more direct, says the same thing, and is easier to decode for the reader.

This is not to say that short sentences are always best; leaving out essential information the reader needs from a sentence undermines its clarity and capacity to convey ideas. There are long sentences made up of phrases and clauses embedded within each other and separated and organized by punctuation marks that convey difficult or nuanced ideas well. Sometimes, short sentences fail.

Wordy phrases

Many technical communication textbooks provide lists of what Mike Markel terms "wordy phrases" that can be replaced with concise phrases. The table shows some of Markel's examples which involve reducing text not from a longer to a shorter phrase, but from a phrase down to one word (Markel 248).

Table 4.1 Phrases that can be reduced to single words

Phrase	Word
at an early date	soon
at this point in time	now
due to the fact that	because
during the course of	during
subsequent to	after
in order to	to
in the event that	if
it is often the case that	often
so as to	to
until such time as	until

There are many more ways to reduce phrases to simpler phrases or single words. For example, "at this point in time" can be reduced to "at this time" or "at this point." Study this list because it shows that many phrases, of all varieties, can be substantially shortened.

We also need to be mindful of the fact that not all phrases need to be shortened. For example, the following phrase appears in the Preamble to the U.S. Constitution:

in Order to form a more perfect Union

If we were to slavishly follow the recommendations in the table, we would ask that this be replaced with:

to form a more perfect Union

Given the political situation at the time the phrase was written, in which people with different views of government were being asked to decide on a new set of rules, which phrase might have been the most effective? Is it always best to reduce a phrase to one word?

Fancy versus simple verbs

Verbs can also be simplified to meet the needs of a wider audience, and so the writer does not come across as someone who is trying to show off his vocabulary. Both Markel (249) and Anderson (279) use the following substitutions for what they both call "fancy" words or verbs:

Table 4.2 "Fancy" verbs and simpler equivalents

Fancy verb	Common verb
ascertain	find out, learn
commence	start, begin
endeavor	try
initiate	begin
terminate	end, stop
utilize	use

Usually, the simpler verb can help make your online writing more accessible. However, if you are writing a report and you keep using the word "show" throughout it, why not change it up occasionally and use a verb such as "illustrate" or "reveal"? For example, "the information shows" can become "the information illustrates" or "the information reveals." The word "illustrates" has a particular power as it suggests that a picture is being drawn so you can better see something. "Reveals" can mean that something that has been hidden from you is now exposed. Sometimes changing up our prose helps keep it vibrant.

There are many more pairs of fancy/simple verbs, and consulting a thesaurus when you revise is the most effective method if you have any doubts.

The inverted pyramid and the F-pattern

The "inverted pyramid" style of writing requires that we place the essential message of the article, news report, or other body of written information at the beginning, and as we work our way through it, each succeeding sentence and paragraph should be supportive but of lesser value. This method has been taught in schools of journalism and is still the model for newspaper writing today because readers want to get to the essence of a story; the facts of a story need to come first, not ideas and chronologies of events (Scanlan).

Jakob Nielsen was the first to assert that writing for the Web should be based on the inverted pyramid technique. Early usability studies done by Nielsen showed that only about 10 percent of users actually scrolled down the screen after selecting a link and concluded that readers are looking for only the essence of the text (*Designing* 112). Additional eye tracking studies by Nielsen show that if one placed the "two horizontal stripes followed by a vertical stripe" of an "F" on a body of text on a screen, this would describe the tracking pattern that readers use when they read something online ("F-Shaped Pattern"). They might read the first few lines, then scan down the left side of the page and pick up a few words at the beginning of each sentence. Most people will only read the top of a page on a Web site and then move on.

One of the problems with the inverted pyramid style of writing is it reduces the complexity of events just to what has happened and not the background, nor does this pared-down style do justice to "stories" that are profound and have a lasting value. Consider this opening to an article by Robert Kaplan (54) on Vietnam:

> The effect of Hanoi is cerebral. What the Vietnamese capital catches in freeze-frame is the process of history itself—not merely as some fatalistic, geographically determined drumroll of dynasties and depredations but as the summation of brave individual acts and nerve-racking calculations. In the city's History Museum, maps, dioramas, and massive gray stelae commemorate anxious Vietnamese resistances against the Chinese Song, Ming, and Qing empires in the 11th, 15th, and 18th centuries. Although Vietnam was integrated into China until the 10th century, its political identity separate from the Middle Kingdom ever since has been something of a miracle—one that no theory of the past can adequately explain.

If we had to summarize this paragraph in the spirit of today's inverted pyramid style of writing, we might reduce it to this sentence:

Hanoi is a city that vividly illustrates its one thousand years of dramatic, turbulent, and enigmatic history.

Furthermore, to capture the overall point Kaplan is making in his article, we could start with the following:

Vietnam has become an ally of the United States to ward off the threat of China.

Nielsen and others want us to present information in digestible or chunked units of easily comprehensible text because this best allows viewers to pinpoint and read what they need and then decide if they should read on. This is certainly important; we need to serve our busy readers so they can find what they need and then make informed decisions. However, the summarizing sentences in this example are arid and formulaic, and do not have the rhetorical power to elevate this essay above other news stories about much less important matters. Still, for most kinds of information that we need to find online, we should steer clear of Kaplan's style of writing.

As opposed to starting a Web site with a single sentence or tidily phrased tagline, as described by Krug (and as discussed in more detail in Chapter 5), an explanatory paragraph can be useful, as it allows the writer to make a case for the value of the topic. This is shown below:

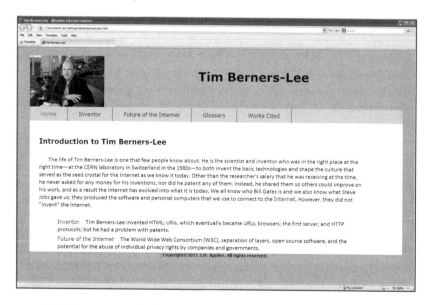

Figure 4.1 Home page and explanatory paragraph.

Of course this could be reduced to something even shorter:

Tim Berners-Lee invented the Internet, took little money for it, and is relatively unknown among the general public today.

As the Kaplan introductory paragraph loses something when pared down to one sentence, this distillation of the description of Tim Berners-Lee's work would lose something too; there would be little concrete background to make his story compelling. While it might satisfy some editors who want everything reduced to one sentence, just stating a few facts does not always work in rhetorical contexts such as this one.

In other rhetorical contexts including less textual content can be effective; for instance the personal Web site example described in Chapter 6 includes just the name of the person at the top of the home page, a simple graphic, and four links that connect to the person's biography, resume, writing samples, and professional links (see Figure 6.2 on page 265). When we come to such a site we can begin to make choices and learn more without an introductory paragraph. When we type in the URL for someone's professional or personal Web site, it is to see if she might be someone we would want to work or be friends with, so simple "biography" or "resume" links provide enough information to get us started.

In contrast, Tim Berners-Lee might be someone we have heard of but we only have an unformed or inchoate sense of his place in recent history, and an introductory paragraph would quickly contextualize this person for us.

Redish (107) refers to large bodies of text as "walls of words" that become "barriers" for users, and argues that we should break them down into lists or short paragraphs when we can. The writing professionals at the Mayo Clinic break down such a wall of words on the clinic's Web site documenting the characteristics of diseases and conditions, the processes of identifying and treating or managing them, and the relevant ongoing research and education efforts of medical professionals. To this end, they organize their documents into some combination of chunk-sized texts in the following categories: Overview, Diagnosis, Treatment, Appointments, Clinical Trials, Research, Symptoms, Causes, Risk Factors, Tests and Diagnosis, Treatments and Drugs, Lifestyle and Home Remedies, Coping and Support, Alternative Medicine, and Complications. These texts are separate, yet linked to one another.

In the following screenshot we have the overview of REM Sleep Behavior Disorder from the Mayo Clinic Web site:

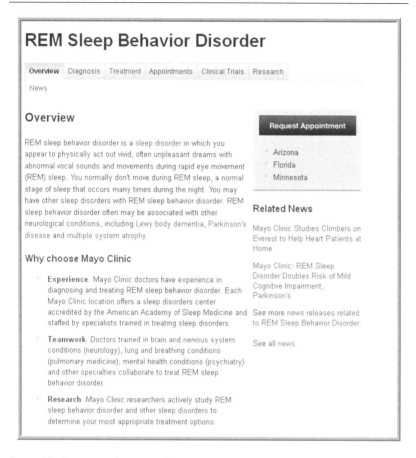

Figure 4.2 Overview information for REM Sleep Behavior Disorder on the Mayo Clinic Web site.

Not all diseases and conditions have overview sections on the site; usually, these are used for those maladies that are complicated and relatively obscure. For example, the common cold does not have an overview section.

The next screenshot shows the "Treatment" text for REM Sleep Behavior Disorder on its own separate tab. Not only does this layout device break down the wall of words, it allows a user to go straight to "Treatment" if he already knows he has this condition:

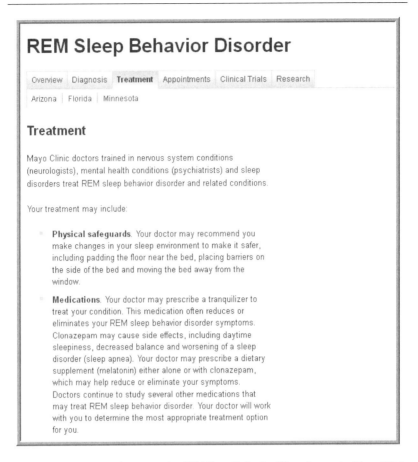

Figure 4.3 Treatment information for REM Sleep Behavior Disorder on the Mayo Clinic Web site.

Had all of the information in the six tabbed categories been packed into one longish page, especially in an online environment, the reader would be facing a large block of writing that would not be as inviting as it is in this hyperlinked form.

We also can eliminate a wall of words by breaking a paragraph down into lists or clearly demarcated smaller sections of writing. This is demonstrated in the following screenshot:

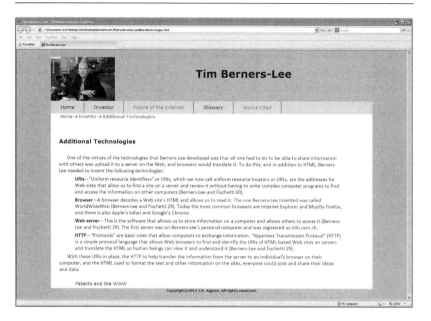

Figure 4.4 Breaking down a paragraph.

The separation of the four technologies described (URIs, Browser, Web server, and HTTP) and the fact that they have been listed in bold makes the work of Tim Berners-Lee more evident than had they been buried in a larger paragraph.

As described by Bolter and others (see Chapter 1), hypertext structures by their very nature challenge the organizational structures of conventional texts. For example, the structure of the writing used to produce the Tim Berners-Lee example could have been presented as a conventional essay with this linear pattern:

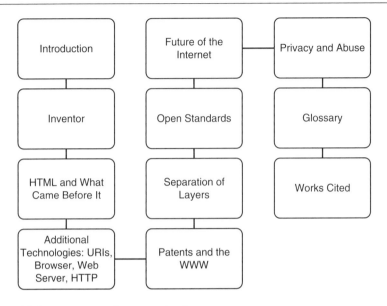

Figure 4.5 Linear pattern of a conventional essay or report.

As you will see, two different hypertext structures are employed for this Web site in Chapter 6, each conveying the same content: one involving a breadcrumbs style of menu, and one a drop-down style of menu. Any time we use devices such as enumerated chapters in a book or headings in conventional print text, we are breaking down the wall of words into smaller units. It is just a matter of degree. Hypertext structures have similar organizing devices. For example, the "Additional Technologies" file for the Tim Berners-Lee Web site could consist of a brief introduction to the HTML files for URIs, Browser, Web server, and HTTP, and serve as a gateway and link to each of these:

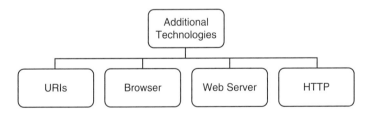

Figure 4.6 Hierarchical pattern of a hypertext structure.

Hierarchical representations, especially large-scale ones, can look daunting and overly complex at first, but if organized thoughtfully they can break down larger bodies of text and make them easier to understand.

Exercises

1. Read an article in a local or a national daily newspaper and see if the reporter has presented a story using the inverted pyramid technique. As you work your way down the article, do you sense that the information is just not very important, or at least much less important? Compare this organizational strategy to that used in a major article in a magazine such as *The New Yorker*, *The Atlantic*, or *Harper's Magazine*. Sometimes in these magazines the most important ideas are presented near the end of articles.
2. Review a passage in any textbook that you have that details a process, presents a set of directions, or defines more than one idea or thing. Would this be a good candidate to be made into a list or set of clearly demarcated smaller texts or paragraphs, as in the example of the "Additional Technologies" file?

Style and punctuation

We use punctuation to more clearly signal the precise meaning of our ideas; punctuation marks allow us to write and express ourselves clearly. Punctuation allows us to alter the rhythm and organization of our sentences and this provides them with greater rhetorical power. Imagine how Thomas Jefferson's Declaration of Independence or Barack Obama's "Race Speech" would sound without the appropriate use of punctuation.

We have to understand that each punctuation mark is a tool or device that has a specific meaning. Sometimes people will insert, for example, a semi-colon in a sentence because they have already used a comma and it just "feels right"—but if they do this without knowing the difference between a semicolon and a comma then it may not have the effect for the reader that they want. In fact, it could lead to a more confusing sentence. Learning the correct use of punctuation marks will keep this from happening.

Learning how to punctuate correctly also gives you greater power as a writer and will serve you well in your career.

Sentences

The following two sentences each have a subject and a predicate.

> Don DeLillo is a novelist who presents thoughtful insights into contemporary culture and technology.

> He often writes about how language and media influence the way we understand ourselves and our relationship to others.

In these sentences, the subjects are either "Don DeLillo" or "He." The predicates are the words that follow the subjects and they describe who

Don DeLillo is or what he does. In the first sentence the predicate is "is a novelist who presents thoughtful insights into contemporary culture and technology." Complete sentences have both a subject and a predicate and are often referred to as "independent clauses"; they can stand independently and present a complete thought.

As written, these sentences are fine, and the second sentence could follow the first in a paragraph. Also, they are complete thoughts. "Fragments" are not complete sentences or complete thoughts. I think we can all agree that "Presents thoughtful insights into contemporary culture and technology" is not a complete thought or a sentence. What does this mean, and where is the subject? It is a fragment.

Semicolons

COMBINING INDEPENDENT CLAUSES

We can also combine these two sentences with a semicolon:

> Don DeLillo is a novelist who presents thoughtful insights into contemporary culture and technology; he often writes about how language and media influence the way we understand ourselves and our relationship to others.

Semicolons are unique; they say that, in the next sentence, the writer is now going to further detail what she or he said in the previous sentence. They draw the readers in so they can better see the connection between both sentences. Oftentimes semicolons are misused; people substitute commas or colons in their place.

Semicolons are best used when the sentences they connect are closely related, as they are in the example directly above; in the second sentence I am illustrating in greater detail just why Don DeLillo offers thoughtful insights into contemporary culture and technology. However, if we connected the following two independent clauses with a semicolon, we would not be using the semicolon to its fullest advantage:

> Don DeLillo lives in New York City; his novels have won many awards.

While we are joining two independent clauses with a semicolon here, living in New York City does not have anything to do with winning literary awards. Sometimes deciding whether or not two sentences are tightly connected is a judgment call; perhaps some New Yorkers think that being an award-winning novelist requires that one also live in New York City, but I feel this reasoning is questionable. I think these two sentences could be placed next to one another in a paragraph, but they should not be connected with a semicolon.

With conjunctive adverbs

Semicolons can also be used to join two independent clauses where the second independent clause begins with a conjunctive adverb. The following are conjunctive adverbs: thus, therefore, certainly, also, consequently, however, otherwise, moreover, furthermore, similarly, still, instead, besides. It is important to realize that conjunctive adverbs are used to tightly connect both independent clauses. Some examples of the use of conjunctive adverbs and semicolons follow.

> Semicolons can also be used to join conjunctive adverbs that are followed by an independent clause; moreover, conjunctive adverbs sit at the beginning of these independent clauses for the purpose of qualifying or setting up all that is to follow.

The conjunctive adverb here is "moreover" and what follows it is the second independent clause, which further clarifies the function of conjunctive adverbs.

> Jennifer wanted to go to the beach; however, Jason wanted to go to the mountains.

"However" is the conjunctive adverb here and it sets up the independent clause that tells us that Jason had different ideas than Jennifer.

> The professors forced the reluctant students to attend a thrilling football game that the home team won in overtime; instead, the students would rather have been in their dorm rooms studying punctuation.

"Instead" is the conjunctive adverb here.

With items in series

Additionally, semicolons can be used to make sentences with items in series read more clearly.

> Don DeLillo lives or has spent time in New York, New York; Athens, Greece; and Tucson, Arizona.

If we used commas for semicolons here, all these place names would not be properly separated:

> Don DeLillo lives or has spent time in New York, New York, Athens, Greece, and Tucson, Arizona.

As you can see, without using semicolons, these cities, states, and country are not properly contextualized relative to one another. I know the reader knows Athens is in Greece, but the use of semicolons still makes it easier to read.

Comma splices

A "comma splice" is a misuse of a comma when what is needed is a semicolon, period, or comma combined with a coordinating conjunction.

> Don DeLillo is a novelist who presents thoughtful insights into contemporary culture and technology, he often writes about how language and media influence the way we understand ourselves and our relationship to others.

Joining the two independent clauses here with just a comma is a comma splice. However, you can use a comma and a coordinating conjunction to remedy this. The following are coordinating conjunctions: for, and, nor, but, or, yet, so. We can get out of the comma splice in the example by writing the following sentence:

> Don DeLillo is a novelist who presents thoughtful insights into contemporary culture and technology, for he often writes about how language and media influence the way we understand ourselves and our relationship to others.

In terms of punctuation, this sentence is perfectly fine.

Colons

A colon is used to introduce a list. I have already used one above in "The following are coordinating conjunctions: for, and, nor, but, or, yet, so."
A semicolon or a comma is not used to introduce a list.

Introductory phrase

An introductory phrase can be used to set up the complete sentence that follows, and, if it is a relatively long introductory phrase, it should be separated from the complete sentence with a comma.

> Before writing his novel *Ratner's Star*, Don DeLillo spent a lot of time studying the history of mathematics.

The introductory phrase is "Before writing his novel *Ratner's Star*." The independent clause or sentence is "Don DeLillo spent a lot of time studying the history of mathematics." An introductory phrase is not an independent clause; it cannot stand alone as a complete sentence or a complete thought. Imagine someone saying to you "Before writing his novel *Ratner's Star*" in a conversation. It would be an incomplete thought.

Commas and adjectives

COORDINATE ADJECTIVES

When we use coordinate adjectives to modify a word, we separate them with a comma:

> The warm, gentle instructor began the yoga session in a soft, friendly voice.

When use commas between the adjectives that modify a word, we can substitute the comma with "and" and it will still sound right:

> The warm and gentle instructor began the yoga session in a soft and friendly voice.

We could also change the order of the adjectives and retain a proper grammatical structure with coordinate adjectives:

> The gentle and warm instructor began the yoga session in a friendly and soft voice.

CUMULATIVE ADJECTIVES

As opposed to offering a string of adjectives that can be thought to modify a word separately and independently of one another, a list of cumulative adjectives that are not separated by commas modify nouns in combination with each other. Perhaps this is why they sound out of place when we invert them.

> Twenty-seven healthy people began the yoga session with some simple warm-up stretches.

If we wrote "Healthy twenty-seven people," it would not sound right. Nor would "Twenty-seven and healthy people." The same goes for the phrase "simple warm-up stretches."
Given what you have learned, why is the following sentence correct?

> The gentle, warm yoga instructor began the session in a friendly, soft voice.

Why do we have a comma between "gentle" and "warm" but not between "warm" and "yoga"?

Often people are told to omit the second comma in a series before the "and," but sometimes this gets you in trouble. When we do this in the following sentence, it is misleading:

> I owe my success to my parents, Hillary Clinton and Tiger Woods.

In this sentence, you are stating that Hillary Clinton and Tiger Woods are your parents, rather than people you think of as role models who inspired you. The following would be better:

> I owe my success to my parents, Hillary Clinton, and Tiger Woods.

If you want to indicate that your parents had something to do with your success and also tell us their names, this construction would work well:

> I owe my success to my parents, John Smith and Mary Jones.

It is best always to use the second comma in series before the "and" unless the relatively unusual sentence construction such as this comes into play. If someone tells you that you should omit the second comma, show him or her these examples.

Exclamation points

Exclamation points should only be used in dialogue in plays and fiction. For example, we might see "'It wasn't me!' he exclaimed" in a short story or play.

Exclamation points are the punctuation equivalent of shouting, and you can say powerful things without shouting if you get the words right, as Jefferson and Obama have. Shouting puts people off.

Possessives and apostrophes

A possessive shows that something belongs to or is associated with a noun, and is designated by an apostrophe followed by "s":

> That is Obratovich's car.
> She was sitting in the driver's seat.
> We are not just talking about your right to free speech, but everyone's right to free speech.

Adding an apostrophe "s" to a word that already ends in an "s" can sometimes look awkward. In this case, just an apostrophe will do:

Dr. Seuss' *The Cat in the Hat* is the only book one needs to read and understand to be the secretary of state.

When writing the possessive form of the pronoun "it" we do not use an apostrophe "s," because this would give us the contractive form "it's" which means "it is":

When we speak of the Bill of Rights, we can be sure that its place in American history is significant.

Dashes

Dashes are ways of making a point or further embellishing an idea within a sentence.

When McLuhan uses the phrase "the medium is the message," he means that to really understand the medium—the technology that we create—we need to understand its larger effects on our culture or society.

Dashes are used here to further clarify just what "medium" means, which for McLuhan is a key distinction. Dashes have a greater emphasis than parentheses, which could have been used in this sentence but would not have been as effective. They should be used sparingly.

On a technical note, dashes are twice as long as hyphens, and, rather than using two hyphens to indicate a dash, holding down the "Alt" key on a keyboard as you enter "0151" will give one contiguous dash.

MLA in-text citation style

There are many good guides on how to use citation styles such as MLA and APA in "Works Cited" or "Bibliography" pages, and these can be easily found online. There are fewer guides on how to incorporate or cite works in text using these styles.

In the passage below, consider how research is incorporated into the writing:

Richard Lanham believes that our responses to writing and graphics on computer screens can be likened to the way we have responded to oral discourse, as opposed to print in books, in the past (34). When we take in information on a screen, "the text becomes unfixed and interactive" and the reader can "change it" and assume the role of writer (31). Lanham tells us that the "electronic word" that we come into contact with on our computer screens affects our perception or "human sensorium" differently than the traditional printed word. Traditional print texts have "No pictures; no color; strict order of left to right then down one line; no type changes;

no interaction; no revision" (34), and thus make the reader feel that the writer's textual meaning is not open to any interaction.

Here is why the citation in this passage works:

1. In the first line the writer summarizes an idea from Lanham and gives him credit. There is no quotation.
2. In the lines that follow, parts of sentences from Lanham's work are quoted, and these are surrounded by the words of the quoting writer, the "connective tissue" of the paragraph. The pieces of Lanham's writing that are quoted pack a punch and would lose their power if they were summarized.
3. The effect is that the quoting writer is having a dialogue of sorts with the quoted writer, Richard Lanham. The quoting writer is not just slapping in quotes or summaries, but instead showing he understands his sources and can use them for his own purposes.

When we use a "blockquote," we are setting a longish quote aside from the text it is embedded in. Usually if a quote is one word over three lines or longer, it should be in blockquote form as we have below:

> "Digital literacy" is another term that is being used by professionals these days. It conflates some of the elements of the concepts described above, but also extends them. A good working definition of digital literacy has been proposed by Rachel Spilka (8):

>> Theory and practice that focus on use of digital technology, including the ability to read, write, and communicate using digital technology, the ability to think critically about digital technology, and consideration of social, cultural, political, and educational values associated with those activities.

> To become digitally literate, we are asked to be writers and communicators who can use digital technologies and be able to "think critically" about them across a broad range of concerns. Understanding the "social, cultural, political, and educational values" requires that we examine how communication technologies present and shape information. It is not about the content digital technologies deliver, but about how the machines themselves actually affect and perhaps alter our understanding of the content. To do this, we need to understand the history of communication technology and how it has been theorized so we can become more critically engaged as we work to produce and present information.

Here is why this passage works:

- There is a full-bodied lead sentence before the quote that tells us about what is coming. This lead does much more than "Spilka says:".

- After the quote, as opposed to moving on to discuss the next idea, the passage beginning "To become digitally literate …" fully explains just what Rachel Spilka is saying. In this way, the quoted passage is fully integrated into the point that the quoting writer is trying to make.
- Specific parts of the Spilka quote, such as the term "think critically," are woven into the follow-up passage to more tightly connect it to the primary, quoted passage.

It is important to carefully read and know the works of others when you do research, and then give them credit. While it might seem like a burden at first, writing in this fashion enhances your ethos as a writer, and sometimes allows you to think more closely about your subject and come up with more ideas.

Below are three basic guidelines for in-text citation and MLA:

1. Always have a lead-up to a quoted sentence, sentences, or parts of sentences. Do not write short and obvious phrases such as "He said:" and then go into the quote. For example:

   Birkerts moves from a technical to a natural analogy to emphasize this point when he writes that the word on a computer screen "floats on the surface like a leaf on a river" and is "less absolute" than "the leaf plucked out and held in the hand" (155–156).

   The lead-up to the first quote contextualizes just what the quoted writer is saying, using a frame of reference that meets the quoting writer's needs.

2. If you do not include the quoted author's name in the lead- or follow-up text to the quote, you need to put the name in the parenthetical element:

   Before oral cultures were studied to any extent and consequently given some respect, many scholars assumed that, as it was so "skillful," Homer's *Odyssey* must have been a written text, not spoken, reflecting their failure to see how "sophisticated" and "reflective" an oral text can be (Ong 56).

   This gives Walter Ong credit for his ideas, even though most of the words being used are the quoting writer's. This helps to inform the reader that the quoting author really understands Ong's work, because he can use his own words to explain it.

3. If you have more than one work by the same author in your Works Cited pages, you need to indicate which it is by putting an abridged title in the parenthetical element:

   Early usability studies done by Nielsen showed that only about 10 percent of the users actually scrolled down the screen after selecting a link and concluded that readers are looking for only the essence of the text (*Designing* 112). Additional eye tracking studies by Nielsen show that if one placed the "two horizontal stripes followed by a vertical stripe" of an "F" on a body of text on a screen, this would describe the tracking pattern that readers use when they read something online ("F-Shaped Pattern").

In this example, we have references to Jakob Nielsen's book, *Designing Web Usability*, and to an article of his from the Web called "F-Shaped Pattern for Reading Web Content." Abridged titles of books do not need to be in quotes but should be italicized, while for shorter documents such as articles the title should be in quotes. Because the second of these sources is from the Web, we do not have to include page numbers in the citation.

MLA works cited page

There are as many rules for "Works Cited" pages as there are different types of sources; sources may be books, journals, magazines, government documents, or reputable online sources. The reason that we use a works cited page is that anyone should be able to read your sources to see whether you have used them correctly. Also, using works cited pages enhances your authority as a writer; in effect it is saying that, if someone does not believe you, she can review your sources. We also use others' works cited pages to advance our own scholarship; if you read a book or article and you come across an article citation, you know where to go to find and read it.

The MLA formats for various types of sources are given below. For further help on citing books, chapters, and articles from print and online sources, use the Purdue Owl: http://owl.english.purdue.edu/owl/.

* Book:

  Last Name, First Name. *Title of Book*. City of Publication: Publisher, Year of Publication. Medium of Publication.

  Bolter, Jay David. *Writing Space: Computers, Hypertext, and the Remediation of Print*. 2nd ed. Mahwah, NJ: Erlbaum, 2001. Print.

  Titles of books should always be in italics. If there is an edition number, add it in after the title as shown.

* Article in a journal, and with more than one author:

  Author(s). "Title of Article." *Title of Journal*. Volume.Issue (Year): Page Range. Medium of Publication.

  Rowlands, I., D. Nicholas, P. Williams, P. Huntington, M. Fieldhouse, B. Gunter, R. Withey, H.R. Jamali, T. Dobrowolski, and C. Tenopir.

"The Google Generation: The Information Behaviour of the Researcher of the Future." *Aslib Proceedings.* 60.4 (2008): 290–310. Print.

The journal title is in italics, and the title of the article is in quotes. Because we have more than one author, we use the "last name, first name or initial(s)" form for the first author, then go to "first name or initial(s), last name" form for the authors that follow. This is true of all works cited listings with multiple authors.

• Article or chapter in an anthology or collection:

Last Name, First Name. "Title of Essay." *Title of Collection.* Ed(s). Editor Name(s). City of Publication: Publisher, Year. Page Range. Medium of Publication.

Plato. "Phaedrus." *The Rhetorical Tradition: Readings from Classical Times to the Present.* Eds. Patricia Bizzell and Bruce Herzberg. Boston: St. Martin's, 1990. 133–143. Print.

Note for this example that Plato only has one name.

• Online article:

Last Name, First Name. "Name of Article." *Name of Newspaper.* Date Published. Medium of Publication. Date Accessed.

Bumiller, Elisabeth. "We Have Met the Enemy and He Is PowerPoint." *New York Times.* April 26, 2010. Web. December 12, 2012.

Wikis as sources for research

There is nothing wrong with going to a wiki if you want to understand the background to an issue or find a fact, as long as you know that what you are reading might have been written by someone who knows little more about the subject than you do. Using wikis and carrying out sound research are two different things.

The good thing about wikis is that they allow us quick access to information wherever we are. They also allow people in regions of the world who do not have access to libraries to be able to look something up if they have access to the Internet. Maintaining libraries, while essential, is expensive. Wikis are inexpensive.

However, you should never use a wiki such as Wikipedia in a works cited page. Here are the reasons:

1. As stated, anyone can go online and add something to a wiki. When you read books and journals that are in libraries, or from reputable academic

or professional news outlets, there is a high likelihood that the material has been verified by fact checkers or written by an expert. This does not mean that there is no misinformation in academic or professional sources, but the probability of this is much lower.

2. Going to wikis or Wikipedia and quoting from them reduces the difficult task of research to a simplistic exercise. When we research we have to sift through information and carefully select what we believe to be the balanced and complete truth. Most things worth researching are difficult to understand and we need to read more than one source. When you use Wikipedia, some readers will think that you did not really do your research, and this will undercut your ethos or authority.

3. Wikipedia is becoming a monolith. If we have a society that relies on one source of information, we suffer. In countries that are controlled by oppressive governments people hunger for sources other than those the government allows them access to. This does not mean that the people who contribute to Wikipedia are not honest, well meaning, and intelligent. It means that free societies can better remain free when their citizens have access to and actually practice acquiring information from more than one source.

Copyrights, fair use, and the public domain

Researchers who produce any kind of document use other sources in their efforts. Using them fairly and legally requires that we understand what the rules of use are. What follows here is, in part, a short summary of Stanford University's *Copyright & Fair Use* Web site, which is based on the work of Richard Stim.

For something to be copyrighted, it has to be "original" and created by one or more authors. If the text you want to use is similar to something that came before it can still be understood as original, even if it is not thought to be of high quality. As long as the author "toils" without explicitly copying from some other source, it can be copyrighted (Stanford University). Copyrighted material can be—among other things—written material, images, or audio recordings.

"Fair use" is the practice of using copyrighted material for the "transformative" purpose of commenting on, criticizing, or parodying the material (Stanford University). This is a protected freedom as it allows us to present our ideas on the work of others without the copyright owner suppressing our opinions.

The best way to ensure that you are falling within the guidelines of fair use is to use only a small portion of the work you are referring to, for example a few lines from a song or poem, or part of a paragraph from a prose text.

Works that are in the "public domain" are owned by the public. Government texts such as the Gettysburg Address are something the public owns; the Gettysburg Address was written by a U.S. president while he was in office and has something to do with our government, history, and culture, and is thus owned by the people.

A copyright of a work published in the United States after 1977 "lasts for the life of the author plus 70 years" before becoming part of the public domain (Stanford University). In addition, all works published before 1923 are now in the public domain. Works published in the period from the beginning of 1923 to the end of 1977 retain their copyright protections for 95 years from their date of publication. If their copyrights have been renewed during this time, they can be extended beyond 95 years. These copyright standards are for works that originated in the United States, and other countries have their own copyright laws.

Facts are not protected by copyrighting. If you use the equation $E = mc^2$, originally set forth by Albert Einstein, this does not mean that his estate has to provide you with a permission, as it is an accepted scientific fact. Similarly, if you mention that Al Gore was U.S. senator for Tennessee, this is a known fact and you do not need to ask permission to use this in your own work.

The following five guidelines will help you ensure that you do not use something from a Web site that would warrant a challenge (Stim 6/2-3):

1. Assume that whatever you find on a Web site is protected. Just because it has been put on the Internet or that a Web site does not explicitly state that its contents are copyrighted does not mean that it is in the public domain.
2. Read the "click to accept" or "clickwrap" agreements that typically come with software and downloads and explain how one can use content such as clip art, and follow the rules set in place.
3. If someone complains that you are using text or images that he or she owns, remove it immediately. Continuing to use the material will damage your case if you end up in court, and may increase the amount of damages you have to pay.
4. Investigate the claim right away and ask for evidence from the person challenging you that he or she in fact does own copyright in the material.
5. If you have any doubts, ask for permission from the owners of the Web site.

Additionally, when you use a source, whether it is from a Web site or a printed work, it needs to be documented on a works cited page.

Works cited

Anderson, Paul V. *Technical Communication: A Reader-Centered Approach*. 7th ed. Boston: Wadsworth, 2010. Print.

Birkerts, Sven. *The Gutenberg Elegies: The Fate of Reading in an Electronic Age*. Boston: Faber and Faber, 1994. Print.

Crowley, Sharon and Debra Hawhee. *Ancient Rhetorics for Contemporary Students*. 4th ed. New York: Longman, 2009. Print.

cummings, e.e. "somewhere I have never travelled, gladly beyond." In *100 Selected Poems*. New York: Grove Press, 1994.

Fleming, Jennifer. *Web Navigation: Designing the User Experience*. Sebastopol, CA: O'Reilly, 2008. Print.

Foucault, Michel. *The Archaeology of Knowledge*. Trans. A.M. Sheridan Smith. New York: Tavistock, 1972. Print.

Gergen, Kenneth. *The Saturated Self: Dilemmas of Identity in Contemporary Life*. New York: Basic, 1992. Print.

Kaplan, Robert. "The Vietnam Solution." *Atlantic Monthly*. 309.5 (2012): 54–62. Print.

Krug, Steven. *Don't Make Me Think: A Common Sense Approach to Web Usability*. 2nd ed. New York: New Riders, 2005. Print.

Markel, Mike. *Technical Communication*. 10th ed. Boston: Bedford/St. Martin's, 2012. Print.

National Park Service. *Yosemite National Park, California*. National Park Service, December 11, 2012. Web. December 12, 2012.

Nielsen, Jakob. *Designing Web Usability: The Practice of Simplicity*. Indianapolis: New Riders, 2000. Print.

Nielsen, Jakob. "F-Shaped Pattern for Reading Web Content." *Alertbox: Current Issues in Web Usability*. useit.com, April 17, 2006. Web. January 10, 2013.

Obama, Barack. "Barack Obama's Speech on Race." *New York Times*. March 18, 2008. Web. January 10, 2013.

Pollan, Michael. *The Omnivore's Dilemma: A Natural History of Four Meals*. New York: Penguin, 2006. Print.

Porter, James. "Intertexuality and the Discourse Community." *Rhetoric Review* 5 (1986): 34–46. Print.

Redish, Janice (Ginny). *Letting Go of the Words: Writing Web Content that Works*. San Francisco: Morgan Kaufmann, 2007. Print.

Scanlan, Chip. "Writing from the Top Down: Pros and Cons of the Inverted Pyramid." *Poynter.org*. The Poynter Institute, June 20, 2003. Web. January 10, 2013.

Stanford University. *Copyright & Fair Use*. Stanford University Libraries and Academic Information Resources, n.d. Web. July 15, 2012.

Stim, Richard. *Getting Permission: How to License & Clear Copyrighted Materials Online & Off*. EBSCOhost, 2000. Web. January 10, 2013.

Vonnegut, Kurt. *Breakfast of Champions*. New York: Dial Press, 1973.

Chapter 5

Layout and organization

Chapter overview

Research and writing are rewarding but time-consuming tasks, and to make sure that this work is presented well we start this chapter by describing how we can choose the best typography for our written texts and construct the most effective layout for our Web sites. In addition, we will describe the organizing strategies that allow us to break down our content into manageable hyperlinked units to ensure that our audience members can find the information they need.

Typography

The technical standards for font sizes and styles have already been explained in Chapter 3, as have the basics of CSS syntax and the way it governs font styles. In this section we will show why basic typography is a key element of Web design and why choosing a "Web-safe" font is important. When we choose a Web-safe font, we are choosing a font that has a greater likelihood of being installed in the browsers that people have on their computers. If a browser does not have a font, the patron will not see it, and either a default browser font, or a second- or third-choice font indicated by you in the CSS coding will be substituted in its place.

Alongside whether a font is Web safe, its legibility is equally important. Lynch and Horton (127) and many other online sources identify Web versions of the Times New Roman, Georgia, Verdana, Arial, and Trebuchet fonts as the most useful. Here they are in their desktop publishing versions:

Times New Roman
Georgia
Verdana
Arial
Trebuchet

You can find lots of stylish fonts that are fun to look at, but if you want people to read what you put online the reader's needs should come first, so legibility and being Web safe are key criteria. Computer screens are still unable to provide the resolution of print, but in time this might change and we will have screens that actually display fonts in a more legible fashion.

When using typography correctly, we are able to establish a visual hierarchy that arranges the material for the user's needs. We know this works in print-based texts when we use first-, second-, and third-level headings that signify which elements of text are nested within others, and we can use this in our online texts too. If we use the HMTL heading tags <h1> to <h6> we can achieve this to some extent, but we cannot tweak the fonts to make them look just how we want. However, as pointed out earlier, CSS allows us to refine our fonts in such a way that they are easy to read and distinguish. As in traditional print media, the variation in font sizes, font styles, and spacing between lines of text on Web sites helps create a visual hierarchy using typography and layout.

Upper and lower case letters

Upper case letters should not be used extensively in text, as they are more difficult to read. We learn to recognize words from the shape of letters, and when they are all the same size they are more difficult to read (Lynch and Horton 131). This is evident in the following sentences:

> Philadelphia is also known as "The City of Brotherly Love."
> PHILADELPHIA IS ALSO KNOWN AS "THE CITY OF BROTHERLY LOVE."

For acronyms such as "WWW," we need upper case letters to reflect the fact that they are indeed acronyms. If used throughout text, however, uppercase letters are thought by many to be insulting because they reflect the ethos of someone who does not respect the audience's ability to read and understand something, so they have to be shouted at. This is also true for exclamation points.

Bold, italics, and underlining

Because of the wide usage of Microsoft Word and other similar word processing software, we often associate bold, italics, and underlining as typographical options with similar effects, as the buttons for them are clustered together in the font toolbar.

Bold text should be used when we want to signal a title or heading. Like upper case letters, using bold in a body of text is like shouting and should only be used to draw attention to something of particular importance.

Italics are a more subtle way of enhancing the meaning of a word in a sentence and can be used to great effect if employed sparingly. Oftentimes, italics can be used to emphasize a word such as "and" to signal a connection between two or more words or concepts:

> People need to both speak clearly *and* listen when they are engaged in conversation.

The italic "*and*" makes it clear that both practices—speaking clearly and listening—are important.

Underlining should not be used in Web writing as it can be confused with links. In the age of the typewriter, underlined text such as Moby Dick signaled to editors that the word or phrase should be put into italics when the manuscript went to a printer. Because with word processing software we can use italics, we do not now need to do this.

Layout

The overall design of a site should use a consistent layout pattern so that, when we move around within it, we know that we are still on the same Web site and have a sense of its structure. While choosing the right typography increases legibility, establishing a consistent layout pattern also contributes to legibility. Horton and Lynch assert that, before the title and other detailed elements of a Web site are even deciphered by a patron, the "repeating patterns" of the basic layout help them to establish just where the information is, thus increasing legibility (120). This can be seen in the screenshot:

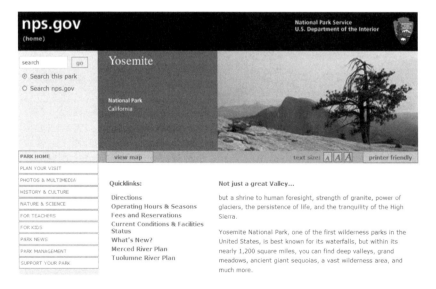

Figure 5.1 Basic layout of National Park Service Web sites.

This home page for a major American national park represents the template for all National Park Service (NPS) Web sites, from Abraham Lincoln Birthplace National Historical Park to Zion National Park. On each of them there is the same spatial layout and the same key links, such as "Plan Your Visit" or "For Teachers," in the left side bar. The situated ethos of the NPS adds to the credibility of the Web site, and the links on the left side bar convey the idea that a visit to a national park can be a meaningful experience and that all are welcome.

However, the "For Teachers" link on this site leads to a page that has layout values that are inconsistent with all of the other pages:

Figure 5.2 Inconsistent layout pattern.

The background color behind the "Interpretation and Education" block departs from the NPS green that we see on all the other pages, and the navigation bar is moved to the right. Additionally, the overall width of the container that encapsulates the other blocks of information is less than on the other pages linked from the NPS Yosemite home page and all other NPS sites.

White space

The use of white space is a practice that draws attention to the most important things on a Web site—the graphics and text—without making the design too busy or cluttered. Lynch and Horton define white space as "negative space," while text and graphics are "positive space" (121). On the opening page of the Yosemite NPS site, the text in the two columns to the right of the side bar are separated by white space. If we had changed the background color or added a border to one of these two columns we would have some more visual separation between the "Quicklinks" and "Not just a great Valley …" sections, but it would perhaps be unneeded and cluttered and we would not be able to use the white background that best promotes ease of reading. It is clear that the middle column serves as a second sidebar of links, and it is easy to distinguish it from the prose to its right. Margins in general can also be used as white space to better separate bodies of text in layout design. The margin to the left of the text in the "Quicklinks" column allows us to more readily identify the links; the top margins in both the second and third columns have a similar role.

"White" space does not always have to be the color white. In the NPS example, there is a lot of space between the text "Yosemite" and "National Park" in the second column, and, while not white, it has the same effect of separating text.

Review the HTML and CSS code in Chapter 3 that describes how to increase the values for margins and padding. This is the most convenient way of managing white space.

Margins and justification

For online documents, left-justified text is the most readable, as the spacing between the words is uniform and the left margin is "even and predictable" (Lynch and Horton 122). The ragged right side of a left-justified paragraph does not undercut its readability.

Right-justified texts can offer an artistic look, but the reader loses the ease with which he can find the beginning of every new line (Lynch and Horton 122). When we are scanning texts, the F-shaped pattern of reading (described in the section on "Writing and Style" in Chapter 4) allows us to understand more readily the work our eyes are required to perform to locate the beginning of every new line. Centered texts, which are ragged at both left and right, are equally difficult to read.

Titles and headings that are placed directly above left-justified texts should also be left justified. Centered headings work well with centre- and right-justified texts but do not work well for left-justified texts, as there is an imbalance. Because left-justified texts work best in online environments, most headings will be centered (Lynch and Horton 122).

The titles, headings, and paragraph text on the NPS pages in the example are all left-justified, and the default layout for HTML coding is left-justified.

Indenting and spacing

An additional use of white space comes into play when we use it to separate paragraphs. This works well if most of the paragraphs are at least two or three sentences. If the text has a lot of short paragraphs, in the form, for example, of lists or quoted dialogue, traditional indenting works well. When we use spaces between paragraphs, as used in the main text of the National Park Service example pages, we get a clean and contemporary look.

The simple non-breaking space code " " can be used for indentation, and the <p> HTML tag will work to make line breaks between paragraphs:

      This is a sentence that is long enough to go across the screen on a contemporary Web site page and it is indented.
<p>This is a sentence that is long enough to go across one screen on a contemporary Web site page and it is encapsulated within two paragraph tags that separate it from the indented paragraph above. This small paragraph is not indented.</p>

Here is how it looks in a browser:

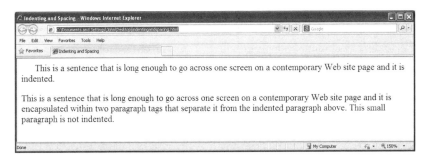

Figure 5.3 Indenting and spacing.

Because there are five   elements, we get a traditional five-space indentation. The <p> HTML tags create a space between the first line and the second paragraph.

The best way to indent, however, is to use CSS. In the example that follows, the code "p {text-indent: 25px;}" is used to provide 25 pixels of indentation, as opposed to inserting five spaces using the   code:

```
<head>
<title>Indent Using CSS</title>
<meta http-equiv="Content-Type"
content="text/html;charset=utf-8" />
<style>
p {text-indent: 25px;}
</style>
</head>

<body>
<p>This is a paragraph long enough to go across the screen on a
contemporary Web page, and it is indented 25 pixels with the use
of CSS code. The code for this is entered between the "style"
tags, which in turn are within the "head" tags. Save this in an
HTML file and see how it looks.</p>
</body>
```

Cut and paste this into a text editor, then save and review it in a browser. Compare it to the previous example.

Screen length

Some files are longer than others and may take up more screen space. Often when we convert a traditional written text into a Web page file we have a tendency to allow the file to run the length of more than one screen, and while this might reflect the initial spatial orientation of the text and graphics it does not necessarily allow us to read it without being disoriented. When you are reading a multiple-page chapter in a book or a long magazine or journal article, you can easily turn the page and reposition your eyes at the top of the next page and know where you are. On a Web site, one long, continuous text does not allow the reader to do this. Depending on the text, the ideal screen length varies, and it is usually between two and four screens.

One possible solution to the disorienting effect of text that takes up many screens and requires a lot of scrolling is to use a "Top" or "Back to top" anchor link. All Web pages that scroll through more than two screens should have "jump to top" buttons (Lynch and Horton 90).

Review how anchor links work in the example of HTML code for the Gettysburg Address in Chapter 2.

Next, create an HTML file using the following code, copying the sentence "All work and no play …" many times, so that the text will extend down for two or more screens:

```
<body>

<a name="backtotop" ></a>All work and no play makes Johnny a
dull boy. All work and no play makes Johnny a dull boy. …

<p><a href="#backtotop">Back to top</a></p>

</body>
```

Now open the file in a browser. Selecting the anchor link "Back to top" should move the viewer to the top of the page.

If you have a long, continuous page, placing a "Back to top" link at screen-length intervals should allow your viewer the sense that he can rise to the surface of the text without drowning in it and becoming disoriented.

Another way to break down a longer text that would otherwise scroll down to an extent that might make the user feel lost is to break it up into separate "pages" that are roughly the same length. This gives the viewer more control, allowing them to move backward and forward almost as if reading a book. Here is the code for "page 1" of a "three-page file" (note that the paragraph text has been elided), which is really three separate files linked together so as to break up the text:

```
<html>

<title>Screen Break with Page Numbers</title>

<head>

<style>
table.right {position: absolute; right: 60px; font-family: Arial;
font-size: 15px;}
</style>

</head>

<body>

<h4 style="font-family: Arial">The Age of Print and the Late Age
of Print</h4>
```

```
<p>A "writing space" is a ...</p>

<p>Bolter's ideas about the effects ...</p>

<table class="right">

<tr>

<td style="padding: 0px 10px 0px 10px"><a href="page2.html">
Page 2</a></td>

<td style="padding: 0px 10px 0px 10px"><a href="page3.html">
Page 3</a></td>

</tr>

</table>

</body>

</html>
```

Below is a screenshot of "page 1." Pages 2 and 3 would have the same coding, with different written content and different links at the foot. For example, the "page 2" file would have links to the HTML files for "Page 1" and "Page 3."

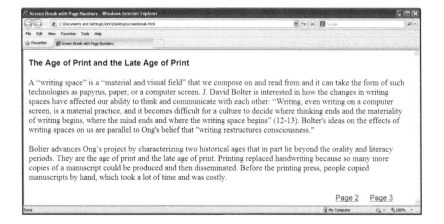

Figure 5.4 Breaking down an extended screen into separate pages.

To position the page number links to the right, we have used basic HTML tags to produce a very simple table of one row with two cells or table divisions. Each table cell has 10 pixels of padding on each side of the text.

To position the table on the right of the browser window, we used a custom CSS class syntax, "right: 60px;". Note that the "class="right"" syntax is embedded in the opening <table> tag. The position is absolute, which means it will be 60 pixels from the right side of the browser window no matter what. We also added some CSS so we would get the Arial font with text height 15 pixels.

Lynch and Horton (90) hold that shorter screen lengths should be used for the following types of Web pages:

- home pages or pages with just menus and navigation structures
- documents that are put in place for browsing or reading
- pages devoted to larger graphics

Longer Web pages are good for the following types of content:

- content that you want to be able to maintain easily
- bodies of text that follow their print equivalents
- texts that need to be easier to print.

Headers and footers

Headers contribute to a site's identity as much as any element in the layout; they are always at the top of the page and the first thing that a viewer will see. Because the efficiency of a Web page is correlated with the number of options in the first four inches of the top of the screen (Lynch and Horton 94), headers need to be seen but not take up too much space. There has to be space for elements such as navigation bars.

The model Web sites in Chapter 6 have headers that meet this requirement and contain only a title block and a simple graphic to provide an identity for the site and establish that all of the separate pages go together. The Mayo Clinic Web site has a simple header with navigation options and a logo that is standardized throughout all of its pages and the simple header that spans NPS pages is also standardized.

Footers establish what Lynch and Horton refer to as the "provenance" or attribution of a Web site (94). On simple sites, the footer indicates in a copyright statement when the site was originally constructed and who it belongs to, and on large-scale sites, other links such as "About …" or "Contact Us" should also appear in the footer area, as shown here: 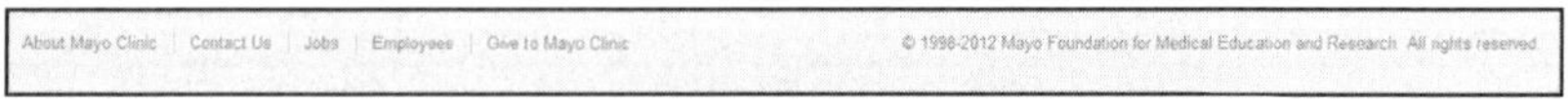

Figure 5.5 Mayo Clinic footer with contact and copyright information.

These footer links are as not as important to viewers as those directly below the header, but need to be in place (Lynch and Horton 94).

Web site architecture

Chapter 1 details in theoretical terms how Web sites are organized differently than traditional print texts and how these differences can affect the way we understand the information online sources convey. In Chapter 4 the basics of organization patterns are presented in terms of classical and contemporary rhetoric. Described in the following sections are some methods that we can use in the organization of Web sites and how the overall body of information in a site can be broken down into smaller units such as home and pathway pages.

Home pages

Ginny Redish holds that the overarching design principle for a Web site is that it should present information as clearly and simply as possible, and we should reduce the amount of words on each page so that readers can find just what they need without having to wade through extraneous information. Regarding the home page of a Web site, she offers some simple guidelines for understanding the audience's needs. People, she says, need to:

- find what they need
- understand what they find
- "act appropriately" when they find what they need (29).

While this is obviously true, applying these guidelines to a Web site is not always easy. The language that labels the navigation elements and home page of a Web site needs to be clear, direct, and provide enough information so that people can "act appropriately," or choose the correct links and move on to find what they are looking for.

For large-scale Web sites built for organizations, she provides a more specific rubric for understanding what a Web site patron needs. The home page of such a Web site needs to:

- identify the Web site and establish the brand
- set the tone and personality of the Web site
- enable people to understand what the Web site is "all about"
- enable patrons to begin locating what they need (Redish 30).

According to Redish, identifying the Web site and establishing the brand requires that the name, logo, and tagline for the organization be in place (31).

"Brand" is a word we hear a lot these days, and it suggests that if a person is a consumer of a certain product or service she can expect to be buying something that has a distinct quality, and can assume that all of the products from the same brand will deliver this same quality. "Branding" can be thought of as the actions an organization takes to get consumers to believe that there is something special about its products.

What does not need to be on a home page is a "paragraph-long mission statement" that explains what the Web site is "all about" (Redish 31), as these paragraphs are rarely read by patrons. Steve Krug more pointedly disparages this "self-congratulatory promotional writing" or "happy talk" on home pages and draws a distinction between writing that says "how great we are" and writing that delineates "what makes us great" (46). This is analogous to the abstract versus concrete language choices described in Chapter 4. If your organization is great, do not say it is great; instead, show the audience members specific information that will allow them to see that it is great. You might have heard one of your writing teachers in the past tell you "Show me, don't tell me," and this is Krug's version of the same principle.

In place of mission statements, Redish believes that taglines are to the point and "memorable," and thus more effective in establishing the Web site (31). Krug describes a good tagline as being:

- clear and informative
- between six and eight words long
- different from others and communicating a clear benefit
- personable, lively, and sometimes clever (103–105).

A more in-depth way to think about taglines is that they are complicated rhetorical entities wrapped in a simple package; they are best understood in terms of ethos, pathos, logos, kairos, and metaphor, as described in Chapter 4. Understanding rhetoric will better allow you to shape a specific and effective tagline for your audience.

Contrary to Redish's formula, perhaps taglines are not always needed, and many Web sites do not use one. Consider the fact that the Ford Motor Company does not use a tagline on its Web site. Its public relations professionals have reasoned that everyone knows that Ford Motor Company builds cars and trucks, so when we land on the company's Web site we just need the name and logo. In contrast, Hyundai Motor Company, which is just establishing its brand in the United States, provides the tagline "Reaching A New Destination By Choosing A Different Path" on its home page. When one selects the image that has this tagline on it, another tagline is revealed: "New Thinking, New Possibilities. By Challenging Convention in Everything We Do, We're Able To Find A Better Way." One can infer that Hyundai Motor Company is challenging the already established automobile manufacturers such as Ford, Toyota, and GM. "A Different Path" is the kind of phrase that can mean many different things and it has a Zen-like quality to it, but in this context it might be used to set Hyundai apart from established firms that have received negative publicity in the last several years: Hyundai will not make the same mistakes as other companies have, because they do things differently. Additionally, Hyundai "challenges" the "conventions" of the larger manufacturers by examining everything they do. "A New Destination" can be thought of as a double entendre, a phrase with a double meaning; when we buy a car we often are buying into the dream and aspiration of driving to new places that will enhance our lives—as opposed to work, school, and the local grocery store—but it can also mean that Hyundai is doing its best to establish itself as a major automobile manufacturer, its "new destination," and this exudes a confidence that might compel consumers to learn more about their products.

Analyzing the rhetorical nuances of the Hyundai taglines reveals a carefully crafted message. Probably the best advice is to realize that taglines for companies that are smaller and not as well established are more important than taglines for established organizations; the smaller company needs an "invented ethos" and the established company can rely on its "situated ethos," rhetorical concepts we discussed in Chapter 4.

Amnesty International's tagline that (at the time of writing) shows up in its Google search result is "Action for Human Rights. Hope for Humanity." What does this tagline suggest? How does it establish this non-profit's brand for us? How might it be setting itself apart from other human rights non-profits such as Human Rights Watch, which uses "Defending Human Rights Worldwide" as its tagline? Are these non-profits in competition, or are they signaling that each organization fills a different niche in the human rights organization field?

When it comes to informational Web sites that teach and explain as opposed to sell something, a paragraph is more appropriate as just a tagline might appear trite and superficial. For example, a paragraph could work well for a Web site describing Tim Berners-Lee—many people already know he is one of a number of information-age industrialists, and a brief paragraph describing what he has accomplished would more effectively enable them to see why learning about him would be to their benefit. For others, just the title "Tim Berners-Lee," a picture, and a few links, however well labeled, might not be enough if they are not familiar with this man. An introductory paragraph is included in the model informational Web site about Tim Berners-Lee that appears in Chapter 6.

Additional elements come into play when establishing the brand of a Web site, and they extend to the site's "personality." According to Redish (31), this is reflected in the site's "visual style," including colors, typography, graphics, words, and writing style. Redish is using the term "personality" here in the same way as many use "ethos" discussed earlier.

Part of communicating what a Web site is "all about," according to Redish, involves providing just enough information on the home page that audience members know what the organization the site represents can do for them; as Redish writes, "What do they have? What can I do here?" (35). Offering too much information can get in the way of our realizing what a site is "all about," so Web site architects have to be careful when constructing the home page or entry point.

The Mayo Clinic home page is rather spare, given the overall size of the Web site, but it does provide a sense of what it is "all about":

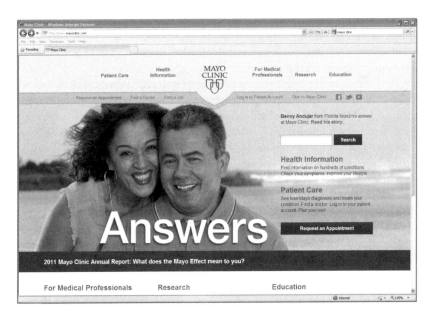

Figure 5.6 Mayo Clinic Web site home page.

The tagline, "Answers," while shorter in length than Krug's recommendation, might be assuring to prospective patients who are confused and worried about their health. It may offer a sense of hope, and the attendant graphic is simple, direct, and human. There is enough information on the opening horizontal navigation bar to direct a patient, with a chance to make an appointment and perhaps get an "answer," and the viewer is invited to read a story about a man who "found his answer."

Regarding branding, the Mayo Clinic can be seen as suggesting that this is the medical establishment you come to when other healthcare organizations have not provided answers. Medical problems can be troubling, and this organization is selling hope. In contrast, some healthcare organizations on their home pages present a picture of the physical buildings they are housed in, monuments to modern medical technology.

A further refinement of the criteria for an efficient home page is "How do I keep going on the question or task that brought me here?" (Redish 36). With this in mind, one principle that would support this effort is that the home page should be mostly links that are short and descriptive, and which direct us to other parts of the Web site. If needed, there should be some text next to each link that explains what it will lead to (Redish 37). As mentioned, "Patient Care" would be enough for most patrons who land on the Mayo Clinic Web site, but there are also options for medical professionals and those who might want to advance their education.

Perhaps it is best to try to choose link titles that do not need an explanation. To amplify this point, home pages on Web sites should provide choices for busy visitors instead of extended descriptions of what the organization's executives want people to believe that the organization is "all about." This brings us back to Redish's succinct questions, "What do they have? What can I do here?" People do not want to read about the mission statement or history of an organization; instead, they want to get to work finding what they need.

One of the techniques now used by institutional Web sites to get patrons on their way to completing their "task" is to provide the forms people need on the home page (Redish 41). While not all Web sites use forms, this practice supports Redish's overall philosophy of simplifying sites to meet people's needs. If possible, forms should be placed as high up on the home page as possible (Redish 41). Additionally, search engine boxes should be at the top of the home page (Redish 43). The Mayo Clinic Web site home page includes a prominent search box, and prospective patients can also select the "Request an Appointment" button and be directed to a form they can fill out.

One final technique for home pages is that the language used to indicate the links should be simple and clear: "use their words—not cute, made-up names that they do not know" (Redish 45). When people are on the Web, they are looking to find the information they need, and they do not come to home pages as experts in the field who might use jargon or in-house phrases. On the Mayo Clinic home page, the links "Find a Doctor" and "Find a Job" work well. "Staff Directory" or "Employment" could signal the same information, but these links lack the same simple active verb structure.

Exercises

1. This section provides a summary of Ginny Redish's ideas about home pages, specifically the four major points regarding branding, personality, what a site is "all about," and how it enables patrons to begin their searches. Pull up any commercial Web site that has wide audience appeal,

and then write about two of these four ideas as they relate to it. How does the home page stand up against these criteria? Make sure you restrict your analysis to one home page. Be specific when you write; as you draw attention to the ideas of Redish and others, make sure you also describe the relevant elements of the Web site and do so in a concrete fashion.

2.   Describe the "discourse community" for patients who are seeking medical assistance. Use Porter's questions from Chapter 4: Who speaks/ writes? To whom do they speak/write? What do they speak/write about? How do they say/write it? How can we use this method of understanding audience to produce a Web site that reflects the needs of this discourse community?

Pathway or first-tier pages

The first set of pages that usually lies one or two links away from the home page and that serves to clarify the basic organizational structure of a Web site are referred to by Ginny Redish as "pathway pages," and their design is of critical importance to a large-scale Web site. While "pathway pages" is a good term for these Web site elements, they can also be referred to as "first-tier pages" and "second-tier pages," because when viewed on an organization chart they are positioned on the first or second level "below" the home page and they usually link to many other pages below that. Using this terminology allows us to imagine how they fit into an entire Web site structure. For example, the links "Patient Care" and "Education" on the Mayo Clinic home page would be pathway or first-tier pages:

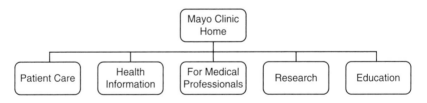

Figure 5.7 Mayo Clinic Web site home page and first-tier hierarchy.

There are often many links on a home page and some contain more information than others. For example, the "Contact Us" or "About" links that we see on so many Web sites technically link to pathway or first-tier pages, but they do not lead us to much more information. The pathway pages that we are interested in are usually listed near the top of a home page or in a position of prominence, and there are probably not too many of them. They are those pages that help break down the vast body of information on a large informational Web site into digestible chunks. If we were to produce a set of links for the U.S. Government home page, we might have three first-tier or pathway pages—Executive Branch, Judicial Branch, and Legislative Branch—as this is the classic organizing structure of American government that we learn in school.

Pathway pages also have sets of links and are dedicated to one area of the overall content of a Web site. They often link directly to pages that contain the information patrons are looking for on large-scale Web sites, pages that do not just contain sets of links (depending on the size of the Web site). Below is a "Patient Care" pathway page from the Mayo Clinic site:

Figure 5.8 Mayo Clinic pathway page.

This pathway page does not present any paragraphs of text giving detailed descriptions of diseases or what is undertaken in the different departments, but it does start the patients on their way. When we break down information that exists in a large, monolithic text into smaller texts that are set out in a well-organized structure, we are "layering" pages (Redish 114). This means that we go from a home page, to a pathway page or series of pathway pages, to the pages with the actual text and images that provide the information the patron seeks.

Redish's term "pathway" is consonant with the terminology of Jared Spool, to whom Redish refers in her work. Spool bases his insights on the idea of the "scent of information," a term he heard from engineers at the

Xerox Palo Alto Research Center, who believe that when people are looking for something on a Web site "they behave like they are hunting for information" (Spool, Perfetti, and Brittan 1). For Redish, your users are seeking a "pathway" to information you might have put in place on the Web. What gets in the way of this need to find information is too much reading (Redish 54), and we can add that there are other things that might get in the way, such as graphics and other texts. In place of a paragraph on a pathway page that tells readers where they have landed, an intuitive set of links should be provided.

These intuitive links, which create a "scent of information," have several identifiable qualities. First, they should have some "trigger words" in them, words that can readily be identified by an audience member and that are relevant to her needs (Spool, Perfetti, and Brittan 2). For example, a student who needs financial support would probably identify the word "financial" and the phrases "financial aid" and "financial assistance" as triggers, and these would be appropriate titles for links. However, a student having a problem with a university policy, one that is perhaps being incorrectly applied to her situation, and cannot find anyone on campus who can help might find it difficult to know what trigger word would be of value, unless she knows what the words "ombudsman," "ombudsperson," or "ombuds" mean. An ombudsperson is an on-campus professional a student would contact if she has a concern about a university policy or practice. Looking for words or phrases like "policy," "university policy," or "problem" using a university Web site's search engine could yield many different kinds of pages other than those associated with the ombudsman. Universities are reluctant to use phrases like "policy problem resolutions," as not all problems can be resolved in an ombuds office, and a link with this phrase might imply this. Link titles need to be carefully constructed, so asking a university ombudsperson for his or her advice on the words or phrases that could be used to define the office and which are also used and understood by students would be a good idea. Ombudspersons have met with concerned students who have explained to them just how they finally found the ombuds office, and this would allow them to suggest trigger words for links.

Sometimes Web sites provide a long list of topics in an index or list of alphabetic links—an A to Z—but Redish argues that these lists can be confusing (56–57). We should add to Redish's concern that, for this strategy to work, people have to know the name of whatever they are looking for; if they do not, an alphabetic list does not help them to start a search. In the case just mentioned, most people do not know what an "ombudsman" is or does, so they would not know to select the "O" link in an A to Z list. However, sometimes a Web site has so much information on it that it probably does not hurt to have an A to Z list. It may provide patrons with other options that allow them to discover, for example, just what an ombudsman is. Perhaps a link on a university pathway page entitled "Understanding university policy" or "Problems with university policy" that connected directly to the ombuds page would complement an A to Z list in which "ombudsman" could be found in the "O"s. Without an A to Z list, those who already know what an ombudsman is might have to work harder to find the right page.

Spool, Perfetti, and Brittan have described research which found that the best links are between seven and twelve words long. The reason that seven words is the lowest value is that, with at least seven words, there is a higher probability that the link will contain a trigger word, as compared to a one- or two-word link (17). As an example, if we apply this to the U.S. Food and Drug Administration Web site we can better see how this plays out. Consider the ten-word link "FDA approves Surfaxin to prevent breathing disorder in premature infants." This could be understood to contain up to eight trigger words, and any combination of two or three—e.g., "Surfaxin," "breathing," and "premature"—would be enough to get some users to the information this page linked to. Just the phrase "FDA approves Surfaxin" might be helpful if the user knows what Surfaxin is. With more than twelve words, trigger words "get lost in the surrounding text" (Spool, Perfetti, and Brittan 17).

While Redish always wants us to eliminate extraneous language on a Web site and asserts that a set of thoughtfully labeled links presented on a pathway page should be enough information, she does point out that sometimes a descriptive sentence attending a link can be useful, if it further describes the link in a way not readily understood through the link's title (59). For example, the FDA Web site has four pathway pages linked from its home page, with the link titles "For Consumers & Patients," "For Health Professionals," "For Scientists and Researchers," and "For Industry." If we go to the pathway page "For Health Professionals," it provides one link entitled "Articles of Interest," with the accompanying description "Articles for health professionals on FDA topics." This is something most people in this audience would understand from just reading the link itself. However, another link on this pathway page entitled "Biologics" has a descriptive sentence that reads "Biological and related products, including blood, vaccines, allergenics, tissues, and other cellular and gene therapies." This is perhaps more useful as it lists the specific "biologics" and does not repeat any language from the link. The question would be, do the majority of health professionals know and agree on what "biologics" are, or does this description provide a "scent" that allows a user in this audience to more easily make a choice about how to proceed on his or her "path"?

One of the myths of Web site architecture that Redish rightfully dispels is the "three-click rule" that says that a user should be able to get wherever she needs by selecting no more than three successive links. It is not about the number of links; it is about the "smoothness of the path" (Redish 63).

If we design a links system with an architecture that is easy to understand, we can use more than three connections for a path to lead a user to the information she needs. This means links should be properly titled, that they should be placed on pages that do not have a lot of extraneous information, and that, if needed, a clear one-sentence description is provided that explains where the link will lead. All of this is based on the assumption that the overall hypertext structure of the Web site is based on the needs of the user, not of professionals that already know all of the information presented on the site.

Redish relies on the work of Steve Krug to further embellish her ideas about how a link structure should work. Krug's whole Web design philosophy is based on the idea that the tiny micro-decisions that we are constantly making in order to move around a Web site are ones we should not have to spend much time on, hence the title of his book, *Don't Make Me Think*. These many decisions add up and can, over time, make our searches exhausting, frustrating, and time consuming.

Krug employs the concept of "satisficing," which is the practice of quickly making a choice that both "satisfies" and "suffices" for the moment. The phrase was coined by Herbert Simon, a Nobel Prize–winning economist, in an attempt to explain consumer behavior. This is an important concept, as it challenges our ideas about how we as consumers behave when presented with options; we do not spend long periods of time rationally analyzing each of our decisions, but instead we make rather quick decisions that "satisfice," or work for our immediate needs (Krug 24). When designing links for Web sites, we need to understand that users are consumers who will not analyze the label on one link versus that on another link; they usually quickly choose one, and if it does not take them to a place they think is useful, we could have done better. Some decisions people make in their lives have real consequences and they tend to think them through carefully, but when selecting links they select them rather quickly, so we need to keep this in mind (Krug 22–23).

Krug seems to have little patience for the navigation schemes that require a user to read extensive directions that, when finally understood, will lead her straight to what she is looking for (26). In fact, he would say users "muddle through" Web sites in a fashion probably determined by the many choices, retreats, and second choices that characterize the "satisficing" style of quick decision making. Muddling through in this way, users do not carefully read directions or quickly survey the overall layout of a Web site; they just figure it out on their own and begin selecting links.

If patrons do "muddle through" Web sites as Krug describes, we should not imagine that we can make a Web site that allows them to immediately understand its overall order shortly after arriving. Spool, Perfetti, and Brittan point out that users only want to get to the part of the site that matters to them, not the entire site (9). They want to find the information they need quickly, so we need to have an intuitive and well-designed Web site. In Spool's terms, we need to start with trigger words that connect the viewer to the content.

The "iceberg syndrome" is a term used by developers to describe the habit of Web site patrons who will not scroll very far below what is initially visible on a screen to look for additional links that will take them along their pathway. Because of this, each link on a Web site should be short. However, Spool, Perfetti, and Brittan have carried out research that challenges this assumption; users will proceed to the bottom half of a page in search of information or links if nothing is in place, such as marketing messages or other extraneous data, that prevents them from doing so (6–7). In fact, longer pages will increase the likelihood that users will keep scrolling and eventually find what they are looking for if the content at the top of the page demonstrates to them that it will be worth it (Spool, Perfetti, and Brittan 18). Short pages, by contrast, often "block" the "scent," because, with a lesser amount of content, users get to the end more quickly without finding what they are looking for and then have to start again elsewhere, which can be frustrating (Spool, Perfetti, and Brittan 18).

Exercises

1. Perform research on a university's ombudsman page, perhaps at the university that you attend. How does this page define what the ombudsman's office does? What might be the trigger words used by an anxious student who needs some help? Using some of these words, write out some link descriptions that are between seven and twelve words long, then identify where on the overall university Web site these links should go.
2. Identify some other words or phrases that, like "ombuds," universities routinely employ but might not be understood as trigger words by students who need some assistance. For example, would a student coming from high school know that she had to go to a place entitled the "College of Arts and Sciences" at a university for some assistance? Just what is a "provost" at a university?
3. Go to an informational Web site that you have used in the past. Perhaps you might go to your bank's site or to the site of a non-profit organization you respect. How long are the pathway pages? How far do you have to scroll down to find information you might need? Is there a sense that each page invites further scrolling, or do you find yourself wanting to hit the "back" button or click back through the breadcrumb links to find what you are looking for? Is there any information on the pathway pages that is extraneous and causes you to move away?
4. Go to a large institutional Web site and identify the major first-tier or pathway page links. Can you identify the site's audience from these

links? Is there a link that would most likely meet your own needs? Is the ethos or persona of this Web site established to some degree by how the links are labeled? Does the Web site seem to welcome visitors?

Organizing strategies

Choosing and then implementing an organizing strategy is perhaps the most difficult thing a Web site architect does; she has to know the content of the Web site and how it might fit together. Additionally, a Web site architect has to take the needs of the audience into consideration; just knowing the information does not mean that she can lead others to it.

Peter Morville and Louis Rosenfeld provided the first complete listing of organizing strategies for Web sites in their book *Information Architecture for the World Wide Web* (26–36). Ginny Redish includes an additional category in her method of organization: "questions people ask" (79). Together, the strategies are:

- alphabetical
- chronological
- geographical
- task-oriented
- topical
- audience-specific
- hybrid
- questions people ask.

Redish's "questions people ask" is consistent with her application of "personas" in her approach to understanding audience.

Known item strategies

The first three categories—alphabetical, chronological, and geographical—are characterized by Morville and Rosenfeld as "exact organization schemes" that allow for "known item searches"; there is no question about where something should be placed in one of these categories (27).

ALPHABETICAL STRATEGIES

Alphabetical searches are used in print, for example in phone books and encyclopedias. The audience member goes to these sources with the name and spelling of the word to hand. On the Mayo Clinic Web site, the names of many "Diseases and Conditions" are listed in an A to Z, and all of the listings serve as links:

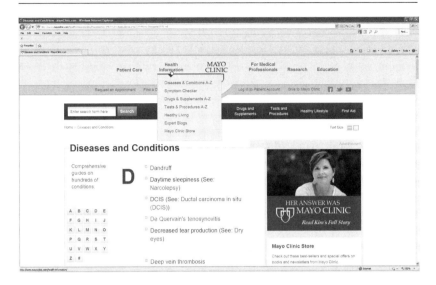

Figure 5.9 Diseases and conditions alphabetic list on the Mayo Clinic Web site.

Most people know what "dandruff" is and would go to the "D"s to see if it was listed, but they might not have heard of "De Quervain's tenosynovitis" unless they had been diagnosed with it by a medical professional. If they knew its spelling, or at least knew it started with a "D" after hearing it, they could go to the "D"s to find and select it. However, if they did not know what it was and had yet to contact a professional about some symptoms, they might find it in the Symptom Checker, a link to which appears in the drop-down menu below the "Health Information" link. Alternatively, if they typed "wrist" and "pain" into the Web site's search engine, De Quervain's tenosynovitis might show up along with several other options.

For a description of how to use HTML and CSS to construct an alphabetical list such as the one on the Mayo Clinic site, review the section on "Alphabetic Links" in Chapter 3.

Some of the advantages of alphabetical listings are that more than one alphabetical index can be used on a Web site, and that there are so many potential applications of them. For example, the Mayo Clinic has four of them accessible through the "Health Information" link: "Diseases and Conditions" and the "Symptoms Checker" have already been described, and there are also "Drugs and Supplements" and "Tests and Procedures." Stemming from the "Research" link are A to Z lists used to organize Mayo Clinic "Research Faculty" profiles and "Areas of Research." The "Patient Care" link leads to "Find a Doctor" and "See Treatment and Diagnosis Services" A to Z listings.

Exercise

Imagine the needs of the students, staff, faculty, and administrators at the university you attend. Will a known item search method such as an alphabetical list work?

CHRONOLOGICAL STRATEGIES

It is easy to understand why history has been taught to us in a sequential manner; what comes before an historical event often informs our understanding of what follows. When we have access to documents that are organized chronologically, this often enhances our understanding of their relevance. For example, when we are searching for information in an academic database or a newspaper archive, the most recent articles and books have most likely taken into account the ideas and facts in the publications that came before them.

Chronological ordering can also enhance the ethos of an organization, demonstrating that it is staying current by consistently offering up the latest information, as in this example:

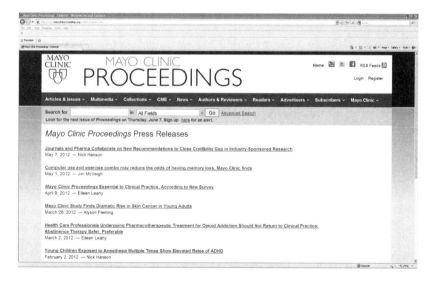

Figure 5.10 Chronological ordering of *Mayo Clinic Proceedings* press releases.

With this chronological listing of press releases, we get the sense that the Mayo Clinic is a dynamic organization in the vanguard of research and serving the medical community responsibly.

GEOGRAPHICAL STRATEGIES

The use of a map as an organizing device has obvious advantages. Maps show us where we are, how we can get to another place, and how places are positioned relative to each other. Many businesses and organizations that are national or international in scale will indicate on their Web sites where we can find a local store or office using geographical methods.

The Mayo Clinic has three separate branches: Phoenix, Arizona; Jacksonville, Florida; and Rochester, Minnesota. A geographical map showing all three places might not be very useful, but a map to find our way around at each location could help. Image maps can be used to show prospective patients the spatial layout of each of the separate clinics, as we see here:

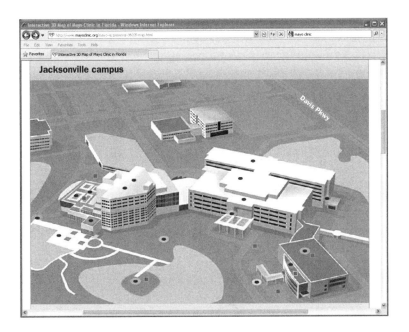

Figure 5.11 Image map of Mayo Clinic, Jacksonville.

Each of the small circles or squares indicates a hyperlink that, when selected, reveals a picture of the particular feature, such as a waiting area, a cafeteria, and the entrance where a patient might need to check in.

For a description of how to use HTML and CSS to produce a map such as the one shown, see the section on "Image Maps" in Chapter 2.

Ambiguous organizing strategies

Morville and Rosenfeld make the point that geographical, alphabetical, and chronological schemes are not ambiguous, but most other strategies are.

How we organize things is subjective; we decide on a set of characteristics, give them names if they do not already have them, and then we use these names to organize something that is abstract into something that is, in our minds at least, concrete. For example, a student's GPA is a characteristic that we often see on a resume, and it has meaning for us: it is a baseline indicator of a student's academic skills. However, this could be challenged. Some students do not take tests well, some students take courses that are easier to do well in than others, and some students have a reservoir of creative, interpersonal, or problem-solving potential that traditional academic methods cannot gauge accurately. However, we have been using GPA as an indicator of a student's potential for so long that we rarely challenge it.

We can argue all day about the words we would use to define a person, animal, place, or thing, but when it comes down to it, we are stuck with them; we need language to convey our ideas to others. But language is often not challenged for its subjective nature. For example, I included "animal" in the traditional "person, place, or thing" list to reflect a different sensibility; animals are "more" than "things." We can parse the real meaning of "thing" all day and never come to a conclusion that would satisfy everyone. The best we can hope for, when we use words to organize, is to understand their subjective nature.

Morville and Rosenfeld have pointed out that, while ambiguity can be frustrating, when we search for things on the Web that are conceptually new to us, we learn how other human beings associate things and concepts (29). Oftentimes we start with an inchoate or vague idea about something and begin plugging words into a search engine, which yields some results that move us closer to what we think we are looking for, and from these results, we can further refine our search with more specific terms, which teaches us just what the terms or words are used by people who are experts on the subject.

TASK-ORIENTED STRATEGIES

Task-oriented sites present audience members with a set of actions that they can take to accomplish an objective. This contrasts with the practice of researching information online; here the audience member is doing something. For example, we can today log into Web sites that allow us to make a purchase, make or cancel an appointment, sign up for a class, or pay a bill.

Using the drop-down menu shown, prospective Mayo Clinic patients are able to perform a number of actions, in addition to logging in:

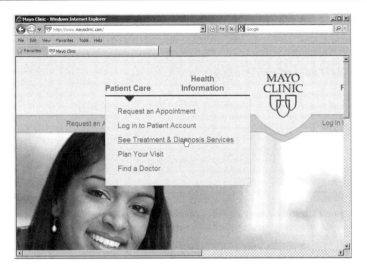

Figure 5.12 Mayo Clinic Patient Care drop-down menu.

We can also make the point that all of the "actions" available in this example start with an active verb: "request," "log in," "see," "plan," and "find." The rhetorical suggestion to patients might be that they can actively begin the process of finding out if they have a medical problem and how the Mayo Clinic might be able to help them. It enables them. Are there any other actions we could imagine being on this list but that were left out for subjective reasons? "Review lab results," "Pay a bill," and "File a complaint" are absent from this list, perhaps for good reason, but a case could be made that they belong under "Patient Care." If the Web becomes increasingly interactive, as some have suggested, we will see how part or all of some Web sites will be designed employing task-oriented schemes.

TOPICAL STRATEGIES

Organizing information on a Web site topically involves identifying the information that it encompasses, breaking it down into parts, and then naming those parts. Hopefully in doing this you capture everything and do not introduce too many overlaps. Morville and Rosenfeld make the point that encyclopedias do this with the "entire breadth of human knowledge" (32), and if you look at the list of the departments at a university you can see how the array of departments, from accounting to zoology, also attempts to do this. However, few Web site architects are tasked with something this grand. Discovering how something understood as a "Healthy Lifestyle" can be broken down would be a more typical challenge, as we see here on the Mayo Clinic Web site:

Figure 5.13 Mayo Clinic Web site—topical organizing strategy.

As mentioned, there may be some overlap. For example, might "Stress management" be an aspect of one's "Fitness" and "Nutrition and healthy eating"? It is important not to undercut an organization's area of expertise by failing to create categories that represent it.

AUDIENCE-SPECIFIC STRATEGIES

When a Web site contains information that attracts distinct audiences, an audience-specific strategy enables more efficient information gathering options to be provided. For example, some universities will have sections devoted to students, staff, and faculty. Staff and faculty members might be more interested in options that deal with the human resources or payroll offices, while students might be interested in links to information on student activities or financial aid. The Mayo Clinic divides the menu bar at the top of its home page into five areas, four of which are based on different audiences:

Figure 5.14 Mayo Clinic Web site—audience-specific organizing strategy.

The "Patient Care" and "For Medical Professionals" audiences are obvious, and there are also links for those engaged in research and prospective students seeking educational opportunities. The "Health Information" link does not specify an audience, but that it is conveniently placed next to the "Patient Care" link and separated from the other audience-specific links by the organization's logo indicates that the primary audience is patients.

Some Web sites could be improved. Jakob Nielsen points out that the needs of users could have been better met by a past version of the United Airlines Web site if its architects had combined the "Mileage Plus" link in the site's "Airline" section with the "Reservations" link in the "Traveler" section. He argued that they should have been combined in a "single top-level category, because they both relate to the individual user's specific data and trips" (171–172). People might want to use their frequent flier miles and make a reservation at the same time. By "top-level," Nielsen means what we have been calling "first-tier."

The HTML and CSS for basic horizontal menu bar structures are detailed in Chapter 3.

QUESTIONS OR FAQS

Redish writes that anticipating questions audience members might have, such as "What's the value of my car?" or "How do I get a new Medicare card?" can serve as a useful guide for structuring Web sites, but notes that the questions that work are ones that people really want answered, "not just ones you want answered" (79). Krug similarly characterizes good questions as ones that people might really have, not "marketing pitches masquerading as facts" (167). No one likes to be manipulated. On the other hand, FAQs, or "frequently asked questions," that really meet the needs of an audience can suggest a positive ethos for the organization that has produced the Web site.

Here is one question and answer from the Mayo Clinic link devoted to nursing school and prospective jobs:

What job opportunities are available for new graduate nurses at Mayo Clinic in Rochester?

Individuals who have graduated from a National League of Nursing (NLN) or Commission on Collegiate Nursing Education (CCNE) accredited bachelors or associate degree nursing program within the last two years may apply for positions in one of more than 60 patient care units or surgical service nursing positions in our hospitals.

New graduates are not considered for Labor/Delivery/OB or the Emergency Department.

New graduates who meet the criteria may be considered for the specialized critical care training program called "Essentials in Progressive and Intensive Care" (EPIC). See <u>orientation</u> for more details.

This question would satisfy both Krug and Redish's criteria for a frequently asked question; a student would want to know where he might be able to apply for a job before going to nursing school. Other questions such as "Does Mayo offer any type of internship or externship for nursing students?" are equally valid. A link to a page devoted to FAQs can be part of a useful organizing strategy for some audiences.

HYBRID SITES

Morville and Rosenfeld point out that "pure" information strategies like the six they describe (as well as the "questions people ask" strategy, as proposed by Redish) allow for easily understood schemes that audience members can grasp and work with as they search for information (34–36), but that organizing large bodies of varied information using one categorization strategy is impracticable. This is apparent in the Mayo Clinic architecture. Imagine using one organization strategy to present this site's information in order to address the needs of audiences differentiated not just by how much each individual audience member knows and needs to know, but also by geographical space. A hybrid scheme is the best alternative, as long as each separate page on a Web site relies primarily on one strategy (Morville and Rosenfeld 36).

Organizational structures

Hierarchical structures

Morville and Rosenfeld point out that the "foundation of almost all good information architectures is a well-designed hierarchy" (37). Their view is refreshing and responsible and challenges the work of others in the field of information architecture. There is a lot of theorizing about the ability of hypertext structures, as opposed to hierarchical structures, to better replicate the associative way that humans put ideas together, but hierarchies, as rigid as they might seem, allow for patrons to form an intellectual template, one that allows them to understand how information has been arranged, albeit by others, so they can find it and use it as needed. Long before hypertext we were using hierarchies such as family trees and charts showing how species are related to organize our worlds (Morville and Rosenfeld 37). Put another way, some things come before others, no matter how you arrange it, and hierarchies are the best way to organize these things.

There are of course problems with thinking too rigidly in terms of hierarchies, as they can set in our minds the idea that one entity is better than another and that power, status, or privilege should be conferred on it. However, if we understand that hierarchies are just intellectual templates designed by some people to help other people find something in a reasonable amount of time, they are the most balanced and practical approach. We use hierarchies, but they are malleable and can be restructured. For example, the content about Tim Berners-Lee used as an example in this book is represented with two different organizational patterns, one using breadcrumbs and another using a drop-down menu.

Hypertext structures

A hypertext document allows for more non-linear connections. For example, in the Berners-Lee breadcrumbs example just mentioned, one has to go through an "Inventor" file to "Additional Technologies" to get to "Patents and the WWW." On a hypertext Web site, the "Patents" page could be accessed from other files in the architecture, such as the "Open Standards" or "Separation of Layers" files, if the material in them allows for a relevant link, thus giving the patron more choices.

Recall the discussion in Chapter 1 of Manovich's challenge to the "myth" that hypertext is really any more interactive than traditional texts: a structure of links can never really capture the unique needs of each user, he argues. Still, adding more links to a hypertext document could be helpful, even if they were put in place by architects with a different perspective on the material than the users.

Navigation technologies

The navigation technologies described in this section are the tools used for moving around a Web site. All of the technologies described can be produced using the HTML and CSS coding examples and exercises provided elsewhere in this book.

Navigation bars

Navigation bars are the most common linking technology on Web sites. As they remain in place regardless of where the user is on the site, they also provide a basic sense of where one is relative to the rest of the site. Usually they are not too intrusive; they are like the controls on a television that go unnoticed until needed, while the user becomes engaged in the content area. They can be used in horizontal and vertical architectures, on the bottom, top, left, or right of the main content area.

Morville and Rosenfeld contend that navigation bars at either the top or bottom, or at both top and bottom of a Web site are more effective than vertical navigation bars. The top navigation bar presents an array of choices on first accessing the page and a basic sense of what is on the site. For sites with longer pages, a set of links at the bottom is appropriate (59).

Drop-down menus

Drop-down menus are extensions of basic navigation bars, and they can neatly package many options for the viewer. However, they can initially hide the options that a site offers because the viewer has to hover over them to see what they reveal. Morville and Rosenfeld point out that, because they have the potential to offer so many options, they need to be used with care (63). In contrast, a basic Web site with many links on the opening page can reveal more. Drop-down menus are best suited to relatively complex Web sites.

Breadcrumbs

When users are able to see the trail of links they have taken to get to their present position, we are using breadcrumbs. As opposed to going "home" or to another link in a menu bar, a user can backtrack one or two steps and restart at a point which is closer to the content he is searching for. Of all navigation technologies, breadcrumbs offer the most clear-cut indication of where a user is and has been on a Web site. They do not appear and disappear like drop-down menus, and in contrast to layers of navigation bars or multiple links on each page, breadcrumbs record with economy the files found in the deep structure of a site.

Krug thinks that breadcrumbs should be supplementary to other navigational technologies, which should be used to show at least the first two levels of the site. He proposes that the last breadcrumb should be in bold, and that breadcrumbs should not do "double duty"—they should not be used in place of a page name, for example (76–79). If we keep breadcrumbs in a small font they are not intrusive, and we can also put a larger heading at the beginning of each page that is identical to the breadcrumb text.

Site maps

Site maps are usually text links provided on one page that reflect in detail the site's overall content. Morville and Rosenfeld emphasize that site maps do not have to present every content file on a Web site to be effective, just the important ones, and this is also true of intelligently conceived tables of contents or indexes (68). If the major sections of a very complicated Web site are laid out and labeled so that they accurately reflect the range of information they contain, this is enough.

In contrast to the first five links on the opening page of the Mayo Clinic Web site, four of which are audience specific, we get much more information here:

Figure 5.15 Mayo Clinic site map.

That there are "Appointments" and "International Visitors" links in the main list here tells us that prospective patients are privileged.

You might find the site map for the Web site of the university you go to or any other large institution you know something about. What have the architects identified as the most important content and what does this say about the intended audience?

Works cited

Amnesty International. Amnesty International, 2012. Web. June 16, 2012.

Ford. Ford Motor Company, 2012. Web. June 15, 2012.

Human Rights Watch. Human Rights Watch, 2012. Web. February 19, 2012.

Hyundai. Hyundai Motor America, 2012. Web. February 19, 2012.

Krug, Steven. *Don't Make Me Think: A Common Sense Approach to Web Usability.* 2nd ed. New York: New Riders, 2005. Print.

Lynch, Patrick J. and Sarah Horton. *Web Style Guide: Basic Design Principles for Creating Web Sites.* 2nd ed. New Haven, CT: Yale University Press, 2002. Print.

Manovich, Lev. *The Language of New Media.* Cambridge, MA: MIT Press, 2002. Print.

Mayo Clinic. Mayo Foundation for Medical Education and Research, 2012. Web. March 15, 2012.

Morville, Peter and Louis Rosenfeld. *Information Architecture for the World Wide Web*. 3rd ed. Sebastopol, CA: O'Reilly & Associates, 2006. Print.

National Park Service. *Yosemite National Park*. National Park Service, U.S. Department of the Interior. Web. June 12, 2012.

Nielsen, Jakob. *Designing Web Usability: The Practice of Simplicity*. Indianapolis, IN: New Riders, 2000. Print.

Porter, James. "Intertextuality and the Discourse Community." *Rhetoric Review* 5 (1986): 34–46. Print.

Redish, Janice (Ginny). *Letting Go of the Words: Writing Web Content that Works*. San Francisco: Morgan Kaufmann, 2007. Print.

Spool, Jared M., Christine Perfetti, and David Brittan. *Designing for the Scent of Information: The Essentials Every Designer Needs to Know about How Users Navigate through Large Web Sites*. North Andover, MA: User Interface Engineering, 2004. Print.

U.S. Food and Drug Administration. *FDA: U.S. Food and Drug Administration*. U.S. Department of Health & Human Services. Web. March 6, 2012.

Major web site projects

Chapter overview

All that we have discussed in the first five chapters of this book supports the two Web site projects presented in this chapter. The personal Web site project will allow you to represent yourself in a favorable and responsible manner. The second project is an informational Web site with two navigation options: breadcrumbs or drop-down menu. Both projects demonstrate how writers can effectively transfer their work to online environments.

Personal web site project

This first project is a personal or professional Web site that allows you to tell others something significant about yourself. Think of this Web site as a calling card that will enable you to get a job, for example, and that every pixel on it will contribute to this goal. Your audience should come away with a feeling that they know who you are. You do not have to use an overly serious tone, but you should imagine that this electronic document might be viewed by a prospective employer.

This Web site will include an opening home page menu file with your name as its title; save it as "home.html." Here is a hierarchical model of the project:

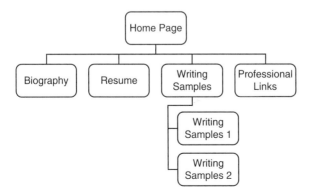

Figure 6.1 Personal Web site hierarchy.

Make sure that you also have your name on the opening menu page, along with a set of relative links to the following elements:

Biography—Write a brief biography, including graphics, that tells us something about you. Think of this as something other than a resume that a prospective employer would look at to find out more about your character. Tell us about your life and interests and include several pictures, placed using some variation of the following CSS code:

.image_left {float: left; margin-top: 5px; margin-right: 8px;}
.image_right {float: right; margin-top: 5px; margin-left: 8px;}

Follow the biography with an image map that allows you to tell us something about yourself. For example you could use an image of the city or state you grew up in and produce a set of links that go to information about your favorite places. You can label the areas on the image map with simple graphics editing software, for example using Photoshop's "Type" tool. Make sure that the image map appears below your written biography, and that you have a lead sentence or two that serves as a transition and tells your viewers what is on the map.

Resume—Use tables to keep the information in your resume in place. As you know how to use colspans and rowspans now, this should not be hard.

Writing Samples—Present at least two papers or creative works that you have written. You need to have a link to each of these texts from the "Writing Samples" link (see the project hierarchy diagram). This makes this page scalable; as time goes on, you can easily add more writing samples. Sometimes it is best to use just the first three paragraphs of a longer paper, so people can get a sense of what you are writing about without being able to use it for their own purposes. The only HTML coding you probably need in the writing samples is paragraph and heading tags.

Professional and Personal Links—List a set of absolute links that tell us something about what you are interested in and deem important. It is best that you have links to professional organizations, and not to things like your favorite political or religious groups. Also, Web sites that suggest that you are interested in things people in your field would also find interesting are important, and a few non-profits that you *sincerely* admire and perhaps support would recommend you as a well-rounded individual. If you have links that only tell us where you go for fun and amusement, this might send the message to prospective employers that you are not focused on your career.

Additional requirements

On the home page and the four other files that stem from it, include a graphic that suggests something about yourself—your taste, your self-image—and that contributes to your site's overall look. Keep this graphic small, as you do not want it to overwhelm the design. Make sure it is a graphic you own. Use a one-row, two-table-divisions layout style. One of the table divisions should be for your name and the other should be for your graphic (see the screenshot of the home page of the sample student Web site presented in the next section).

Employ one external style sheet, use a background that is not too dark and a font color that makes your writing easy to read, and have a "Home" link on each relative link from your home page.

Use some metatags between the <head></head> tags that indicate you are the author of your Web site. This will tell people who view it that you did the coding, and will recommend you as someone who has significant digital literacy skills.

Most importantly, keep your Web site simple and present yourself clearly; it is the information that will get you a job, not garish colors, superfluous links, or wild graphics.

Sample student web site

The following screenshots and HTML/CSS coding are from a Web site produced by Brittney Adams, who has given her permission for their inclusion in this book.

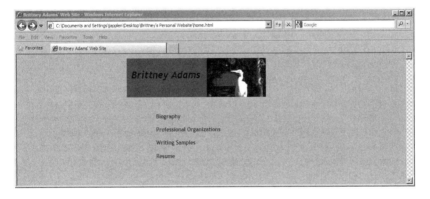

Figure 6.2 Student home page with first-tier links.

This opening page is simple and allows the viewer to begin making some choices about what to review. There is no "Hello, welcome to my Web site" message, just four clearly defined links. Because it is uncluttered, it does not get in the way of what is most important: the writer. Also, it allows the audience members to choose what they want to view, thus they are given some sense of control. In this way, you are meeting the rhetorical needs of your audience.

Here is the code for this home page:

```html
<html>

<title>Brittney Adams' Web Site</title>

<head>
<link rel="stylesheet" type="text/css" href="bladams.css" />
<meta name="author" content="Brittney Adams" />
</head>

<body bgcolor="C3C3E5">
<table width="70%" border="0" cellspacing="0" cellpadding="0">
<tr>
<td width="30%"></td>
<td width="30%" bgcolor="8C489F">
<div style="text-align:center"><h1><i>Brittney Adams</i>
</h1></div>
</td>
<td><img src="bird.jpg" height="127" width="200" alt="Bird"/>
</td>
</tr>
<tr>
<td width="40%"></td>
<td colspan="2" style="padding:50px 0px 0px 100px">
<h3>
<a href="adamsbiography.html">Biography</a>
</h3>
<h3>
<a href="organizations.html">Professional Organizations</a>
</h3>
<h3>
<a href="writingsamples.html">Writing Samples</a>
</h3>
<h3>
<a href="resume.html">Resume</a>
</h3>
</td>
</tr>
</table>
</body>

</html>
```

Note how the writer uses a simple table design and the code colspan="2". This allows the four links below the one-row, two-table-divisions block, containing the student's name and a graphic, to "span" the two columns.

The biography has everything included in the personal Web site project description:

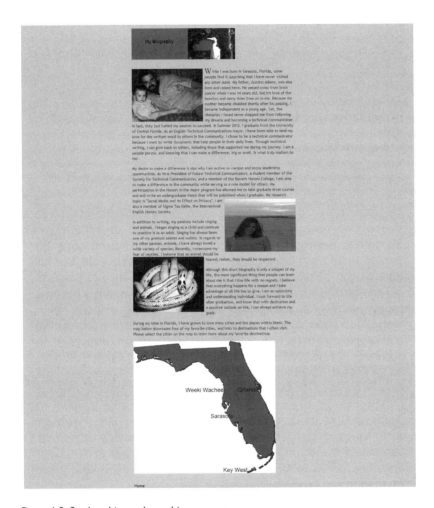

Figure 6.3 Student biography and image map.

Here we have a compelling biography with some graphics that tell us more about the author's life and character. This is the HTML and CSS coding for this page:

```
<body bgcolor="C3C3E5">
<table border="0" cellspacing="0" cellpadding="0" width="70%">
<tr>
<td width="20%"></td>
<td bgcolor="8C489F"><div style="text-align:center"><h3><font
face="Arial, Helvetica" color="#000000">My Biography</font>
</h3></div></td>
<td><img src="bird.jpg" height="127" width="200" alt="Bird"/>
</td>
</tr>
<tr><td> </td></tr>
<tr><td> </td></tr>
<tr>
<td width="40%"></td>
<td colspan="2">
<img src="dad.jpg" height="214" width="300" border="0"
class="image_left" alt="Dad and I"/>
<p class="dropcap">While I was born in Sarasota, Florida, some
people find it surprising that I have never visited any other state.
My father, Gordon Adams, was also born and raised here. He
passed away from brain cancer when I was 14 years old, but his
love of the beaches and sunny skies lives on in me. Because
my mother became disabled shortly after his passing, I became
independent at a young age. Yet, the obstacles I faced never
stopped me from following my dreams and becoming a technical
communicator. In fact, they just fueled my passion to succeed. In
Summer 2012, I graduate from the University of Central Florida.
As an English-Technical Communications major, I have been able
to lend my love for the written word to others in the community.
I chose to be a technical communicator because I want to
write documents that help people in their daily lives. Through
technical writing, I can give back to others, including those that
supported me during my journey. I am a people person, and
knowing that I can make a difference, big or small, is what truly
matters to me.</p>
<p>My desire to make a difference is also why I am active on
campus and enjoy leadership opportunities. As Vice President
of Future Technical Communicators, a student member of
the Society for Technical Communication, and a member of the
Burnett Honors College, I am able to make a difference in
the community while serving as a role model for others. My par-
ticipation in the Honors in the Major program has allowed me to
```

take graduate level courses and <img src="Brit.jpg" height="214" width="300" border="0" class="image_right" alt="Brit"/> and write an undergraduate thesis that will be published when I graduate. My research topic is "Social Media and its Effect on Privacy". I am also a member of Sigma Tau Delta, the International English Honors Society. </p>

<p>In addition to writing, my passions include singing and animals. I began singing as a child and continue to practice it as an adult. Singing has always been one of my greatest talents and outlets. In regards to my other passion, animals, I have always loved a wide variety of species. Recently, I overcame my fear of reptiles. <img src="snake.jpg" height="224" width="300" border = "0" class="image_left" alt="Snake"/> I believe that no animal should be feared; rather, they should be respected. </p>

<p> Although this short biography is only a snippet of my life, the most significant thing that people can learn about me is that I live life with no regrets. I believe that everything happens for a reason and I take advantage of all life has to give. I am an optimistic and understanding individual. I look forward to life after graduation, and know that with dedication and a positive outlook on life, I can always achieve my goals. </p>

<p>During my time in Florida, I have grown to love many cities and the places within them. The map below showcases four of my favorite cities, and links to destinations that I often visit. Please select the cities on the map to learn more about my favorite destinations. </p>

<map name="florida" id="florida">
<area alt="Hemingway Home" shape="rect" coords="480, 534, 505, 521" href="http://www.hemingwayhome.com" />
<area alt="Selby Botanical Gardens" shape="rect" coords="398, 319, 411, 290" href="http://www.selby.org" />
<area alt="Mermaids" shape="rect" coords="371, 213, 395, 191" href="http://weekiwachee.com/mermaids/mermaid-shows. html" />
<area alt="Falafel" shape="rect" coords="470, 215, 494, 187" href="http://www.falafelcafe.com/" />
</map>
<img src="florida.jpg" alt="Florida Favorites" usemap="#florida" width="600" height="537"border="0" class="image_left"/></td>
</td>

```
</tr>
<tr><td bgcolor="C3C3E5"> </td> </tr>
<tr><td width="30%"></td>
<td><a href="home.html"><p>Home</p></a></td></tr>
</table>
</body>
</html>
```

Note how the writer integrates the three images into the text of her biography with the class="image_left" and class="image_right" code. Where the <img src> HTML code is placed determines where the images are positioned relative to the written text. The image map that follows is simple and allows the writer to further illustrate what is interesting to her about the state she has grown up in.

The following resume nicely complements the writer's biography:

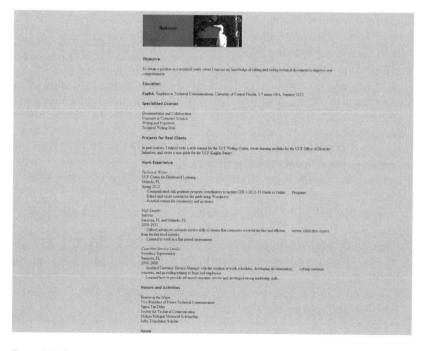

Figure 6.4 Single-column student resume.

This resume covers the student's whole academic career.

Here is the code for this resume:

```
<body bgcolor="C3C3E5">

<table border="0" cellspacing="0" cellpadding="0" width="80%">

<tr>
<td width="20%"></td>
<td bgcolor="8C489F"><div style="text-align:center"><h3><font
face="Arial, Helvetica" color="#000000">Resume</font></h3>
</div></td>
<td><img src="bird.jpg" height="127" width="200" alt="Bird"/>
</td>
</tr>

<tr><td> </td></tr>
<tr><td> </td></tr>
<tr>
<td width="40%"></td>
<td colspan="2">

<h4>Objective</h4>
To obtain a position as a technical writer where I can use my
knowledge of editing and writing technical documents to improve
user comprehension.

<h4>Education</h4>
<b>English</b>, Emphasis in Technical Communications,
University of Central Florida, 3.7 major GPA, Summer 2012

<h4>Specialized Courses</h4>
Documentation and Collaboration
<br>Concepts in Computer Science
<br>Writing and Hypertext
<br>Technical Writing Style

<h4>Projects for Real Clients</h4>
In past courses, I helped write a style manual for the UCF Writing
Center, wrote learning modules for the UCF Office of Diversity
Initiatives, and wrote a user guide for the UCF Knights Pantry.

<h4>Work Experience</h4>
```

```
<i>Technical Writer</i>
<br>UCF Center for Distributed Learning
<br>Orlando, FL
<br>Spring 2012
<br>    -Communicated with graduate
program coordinators to update CDL's 2012-13 Guide to Online
Programs
<br>    -Edited and wrote content for
the guide using Wordpress.
<br>    -Proofed content for consistency
and accuracy.
<br>
<br>
<i>Shift Leader</i>
<br>Subway
<br>Sarasota, FL and Orlando, FL
<br>2009-2011
<br>    -Utilized advanced customer
service skills to ensure that customers received the fast and effi-
cient service which they expect from the fast food industry.
<br>    -Learned to work in a fast
paced environment.
<br>
<br>
<i>Customer Service Leader</i>
<br>Sweetbay Supermarket
<br>Sarasota, FL
<br>2006-2008
<br>    -Assisted Customer Service
Manager with the creation of work schedules, developing docu-
mentation, solving customer concerns, and providing training to
front end employees.
<br>    -Learned how to provide
advanced customer service and developed strong leadership
skills.

<h4>Honors and Activities</h4>
Honors in the Major
<br>Vice President of Future Technical Communicators
<br>Sigma Tau Delta
<br>Society for Technical Communication
<br>Melissa Pellegrin Memorial Scholarship
```

```
<br>Selby Foundation Scholar
</td></tr>
<tr><td bgcolor="C3C3E5"> </td> </tr>
<tr><td width="30%"></td>
<td><a href="home.html"><p>Home</p></a></td></tr>

</table>

</body>

</html>
```

As in the biography, the author uses the colspan="2" code. Once the table is set up, all she needs are tags for headings, breaks, italics, and the   special character.

The following is the professional links portion of the Web site:

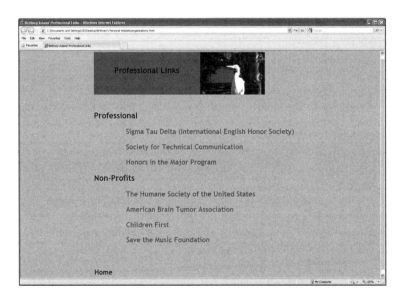

Figure 6.5 Student professional links.

These links demonstrate that the writer is fully engaged in her major and future profession, and the non-profits show her compassion, thus enhancing her ethos.

The following code allows the writer to organize her links:

```
<body bgcolor="C3C3E5">

<table width="60%" border="0" cellspacing="0" cellpadding="0"
align="center">

<tr><td bgcolor="8C489F"><div style="text-align:center">
<h3> Professional Links</h3></div></td>
<td><img src="bird.jpg" height="127" width=200" alt="Bird"/>
</td></tr>
<tr><td> </td></tr>
<tr><td> </td></tr>
<tr>
<td colspan="2">

<h3>Professional</h3>
<p style="padding:0px 0px 0px 100px"><a href=http://www.
english.org/>Sigma Tau Delta (International English Honor
Society)</a></p>
<p style="padding:0px 0px 0px 100px"><a href=http://www.stc.
org/>Society for Technical Communication</a></p>
<p style="padding:0px  0px  0px  100px"><a  href=http://www.
research.honors.ucf.edu/Honors_in_the_Major/>Honors  in  the
Major Program</a></p>

<h3>Non-Profits</h3>
<p style="padding:0px 0px 0px 100px"><a href=http://www.
humanesociety.org/>The Humane Society of the United States
</a></p>
<p style="padding:0px 0px 0px 100px"><a href=http://www.
abta.org/>American Brain Tumor Association</a></p>
<p style="padding:0px 0px 0px 100px"><a href=http://www.
childrenfirst.net/>Children First</a></p>
<p style="padding:0px 0px 0px 100px"><a href=http://www.
vh1savethemusic.com/>Save the Music Foundation</a></p>
</td>
</tr>

<tr><td bgcolor="C3C3E5"> </td></tr>
<tr><td bgcolor="C3C3E5"> </td></tr>
<tr><td bgcolor="C3C3E5"> </td></tr>
```

```
<td><a href="home.html"><p>Home</p></a></td></tr>

</table>

</body>

</html>
```

Review how the padding value of 100 pixels allows the writer to indent her links relative to the headings.

Here is a screenshot of the writing samples page:

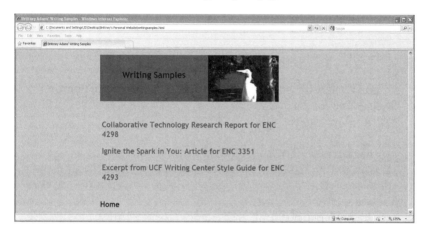

Figure 6.6 Links to writing samples.

The writing assignments are an online portfolio, and the links clearly define the writing samples.

Here is the HTML/CSS coding for the writing samples page:

```
<body bgcolor="C3C3E5">

<table width="70%" border="0" cellspacing="0" cellpadding="0">
<tr>
<td width="30%"></td>
<td width="40%" bgcolor="8C489F">
<div style="text-align:center"><h3>Writing Samples</h3></div>
</td>
```

```
<td><img src="bird.jpg" height="127" width="200" alt="Bird"/>
</td>
</tr>
<tr>
<td width="30%"></td>
<td colspan="2" style="padding:50px 0px 0px 5px">
<h4>
<a href="Collab.html">Collaborative Technology Research Report
for ENC 4298</a>
</h4>
<h4>
<a href="http://writingandrhetoric.cah.ucf.edu/imprint/2011/11
/18/ignite-the-spark-in-you/">Ignite the Spark in You: Article for
ENC 3351</a>
</h4>
<h4>
<h4>
<a href="clientbased.html">Excerpt from UCF Writing Center
Style Guide for ENC 4293</a>
</td>
</tr>
<tr><td bgcolor="C3C3E5"> </td></tr>
<tr><td width="30%"></td><tr><td bgcolor="C3C3E5"> 
</td> </tr>
<tr><td width="30%"></td>
<td><a href="home.html"><p>Home</p></a></td></tr>
</table>

</body>

</html>
```

Review how the padding in the colspan="2" tag positions these links within each table division.

The following is the CSS code for this project, entitled "bladams.css":

```
h1, h2, h3, h4, h5, h6 {font-family: Trebuchet MS, Tahoma,
Geneva, sans-serif;}

p {font-family: Trebuchet MS, Tahoma, Geneva, sans-serif;
color: #000000;}

p.dropcap:first-letter {font: 200% times; float: left;
margin-right: 5px;}

.image_left {float: left; margin-top: 5x; margin-right: 8px;}
.image_right {float: right; margin-top: 5px; margin-left: 8px;}

a:link {font-weight: bold; text-decoration: none; color: #6633FF;}
a:visited {font-weight: bold; text-decoration: none;
color: #330099;}
a:hover {text-decoration: underline; color: #9685ba;}
```

Go back through Chapter 3 and review this specific CSS coding.

Informational web site project

The objective of this project is to reformat a major written research paper into a Web site that effectively conveys all of the essential information in the paper. This can be a report on just about anything, and if you do not have a subject, perhaps something that you are interested in and how it intersects with the World Wide Web would work for you. For example, you can write about online collaborative writing projects in technical communication, or the concerns that some have about the Stop Online Piracy Act (SOPA). General areas of research could include how organizations in education, industry, or government, or non-profits could extend their use of the Internet.

Because we are moving from the traditional medium of print to a hyper-text medium, this project will challenge you to think about the strengths and weaknesses of Web sites. Just coding your research paper into HTML/CSS is not enough to produce an effective Web site. Making it rhetorically effective and easy for the reader to navigate is what is most important.

Overall structure and links—breadcrumbs version

Navigation structure—We have already worked with a navigation structure using HTML/CSS coding. In it we use the classic HTML/CSS structure with a container, header, horizontal navigation bar at the top of the page, a main content area, a footer, and several "Tier 1" and "Tier 2" links. The opening home page, with the file name "home. html," is what our viewers will first see, before they move on to the pathway pages. There should be a "Glossary" page, and a "Works Cited" page listing the sources cited in your paper. You will also need an image for your header. You can use the Tim Berners-Lee Web site referenced in this section as a guide. Make sure you understand how all of the CSS on it works so you can better manipulate it for your own needs.

Home page—Include information on this opening page that is conso- nant with Redish's concept of layering. You should have just enough content here to allow your audience members to understand what your Web site is about, then include links from this opening page to your pathway pages, which will reside on your second tier. This does not mean that you need a huge introduction to your Web site; there should be no "welcomes" or deep corporate philosophy here. However, a short introduction should be included on your home page and there should be several links to pathway pages.

The Mayo Clinic Web site home page is pretty lean in this regard, yet it offers plenty of choices. It opens with options to go to more spe- cific information. Remember that whatever your patrons click on, there will always be links in the navigation structure that lead to some general areas of concern. You can see how the Mayo Clinic site does this with an unchanging horizontal bar at the top of the pages, containing links to areas such as "Diseases & Conditions A–Z" and "Drugs & Supplements A–Z."

Make sure that you keep each linked page that you construct and graphic you might employ in the same folder or file as your home page, or construct appropriate pathways to separate folders with these items in them (see the section on "Files, Folders, and Pathways" in Chapter 2). If you save them in separate folders, they will not be able to link. Name your home page file "home.html."

First- and second-tier links—Employ other links that extend "below" your first-tier links if you need to further break down your information. Make sure that you separate the information into first- and second-tier links in a way that follows the natural flow of your paper. As a general rule, each page should be no more than four computer screens in length. You might also have third-tier links, depending on the structure of your written paper.

Use the breadcrumbs linking style discussed in Chapter 3.

Glossary—You also need to have a glossary that defines at least five key terms and concepts that your audience might need to understand. Use text links with HTML anchor tags to direct the reader to this glossary, as in the Gettysburg Address exercise in Chapter 2 (see the section on

"Anchor Links").Your glossary will be a third-tier link, something that is "below" yet connected to second-tier links. Make sure that you also put a link to this page on the horizontal navigation bar.

Absolute links—Have at least three "absolute" or "remote" links to other Web sites so that your viewers can view information on the Web that is relevant to your subject. They can be to an electronic source that you used for your research which is on your Works Cited page, or included in your Glossary, like the example in Chapter 2 (in the section on "Anchor Links") that goes to the Gettysburg National Military Park Web site. Additionally, you should have an image link to an absolute source. For an example, see the W3C logo included on the Tim Berners-Lee Web site presented later in this chapter.

Graphics—Have at least three graphics in your main content that your text wraps around. Check whether you need permission to use these graphics (see the section on "Copyrights, Fair Use, and the Public Domain" in Chapter 4). Make sure you have a caption for each one, and include any required credit or copyright line.

Pull quotes—Use at least three "pull quotes." These are described in the section on "Additional HTML/CSS Code" later in this chapter. Usually a quote or part of a quote works best for these layout devices.

Works Cited—You should also have a Works Cited page that lists the texts you used in your research. It will look just like the references page of your paper. Include a link to Works Cited in the site header.You need at least four sources in the Works Cited list, and absolute links can also be made from this page. Use hanging indents for each reference (see the Tim Berners-Lee Web site for an example). The CSS code needed to create hanging indents is presented in the "Additional HTML/CSS Code" section.

Metadata—Put some metadata tags between the <head></head> tags that tell the viewer about the author and the copyright information (see the section on "Metatags in Web Sites" in Chapter 3 for more information).

Typography, Graphics, and General Aesthetics

Follow these general guidelines for layout and design:

- Embed text links in your prose; stay away from the infamous "Click here" text link. For example, "You can use W3C Validation tools for both CSS and HTML coding" is much more desirable than "Click here to go to the W3C Web site and validate your CSS and HTML coding."
- Use vertical and horizontal white space to enhance readability, not boxes with thick borders or horizontal lines (rules).

- The only time you should use bold is for headings (if you decide to use headings).
- Limit italics and do not use underline.
- Only use graphics that work for you; superfluous graphics clutter up your page. Make sure that your text wraps around these graphics and that you have a caption for each one.
- Avoid garish color combinations for text and backgrounds; make your work easy on the eye. Research shows that the best background for written texts is white, and the best color for text is black. Save your color accents for your navigation structure and graphics.
- Making your page look distinctive is one thing, but make sure that it looks clean and businesslike. Do not let too many fancy HTML/CSS tricks get in the way of the information.
- Give each screen in your Web site the same stylistic look. This consistency allows your viewers to feel that they have not left your Web site when moving around within it. Putting the same image in your header will help you establish this consistency.

Exercise

Make a flow chart for your Web site that indicates how you are going to break down the information—this is analogous to an outline you might have done in the past for a major written project or group project. Indicate the connections between relative links with solid lines. For absolute links, use dotted lines. You should also include your Glossary and Works Cited pages. Remember that you are working within a navigation structure. This is the Tim Berners-Lee (TBL) Web site structure, without absolute links:

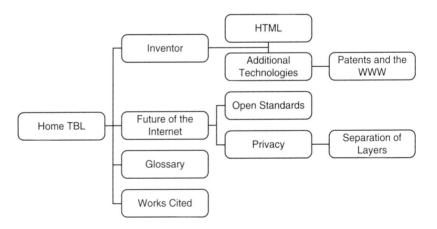

Figure 6.7 Tim Berners-Lee Web site hierarchy.

Method for producing your web site

1. Begin producing the overall navigational structure after figuring out exactly what parts of your paper will be going into each separate file on your Web site. It helps to have done the flow chart exercise above. This first task includes naming all of your files and establishing the horizontal navigation bar at the top of the page, also setting up the breadcrumbs.

    Note: Once you get your first file set up, you can use it as a template to produce the multiple files that will make up your Web site. To put it another way, if you start making your Web site by creating individual pages, then later decide that you are going to reorganize it, you will have to go back and change the organization scheme on a number of these files, not just one. It is important that you get this right at the beginning; it will allow you to put your breadcrumb links in place just once and save a lot of time and energy.
2. Go on the Web to find some images, or use images of your own. Find one image that you can use for your header bar. Ensure you have permission to use graphics that you do not own, and include any credit lines or copyright information as required by the terms of the permission.
3. Produce your "home.html" file, including a brief introduction to the rest of your Web site. In a short space, say what your Web site is about and why the information it conveys is important. Include some short descriptions of the links, as shown on the TBL Web site example later in this chapter.
4. Produce one additional page, other than the home page, including several elements such as images and callouts just to get the hang of it.
5. Manipulate the CSS values to get the look you want on your Web site.
6. Begin producing your other pages by placing the text from your written work into them, along with images, anchor links, and callouts.
7. Identify the terms in the body of your Web site that will find their way into your glossary. Begin writing definitions for them and setting them up with anchor links.
8. Begin your Works Cited page. You can use MLA, or any other citation style required in your field or major.
9. Input your metadata between the <head></head> tags of your home page.

Additional HTML/CSS code

You have seen variations on most of the code presented here in Chapter 3, in the section entitled "CSS Classes." In this section, we are going to detail some of the CSS code that is essential for the informational Web site project, and you will need to understand it to transfer your text from a print to hypertext form. The code presented supports the written word, and writers who want to see their work on the World Wide Web should understand it.

CSS and paragraphs

The CSS code below designates the font family, text indent, margins, and font size for paragraph text:

> p {font-family: Calibri, Optima, Arial, sans-serif; text-indent: 25px; margin: 5px 30px 0px 0px; font-size: 1.0em;}

Note that the default paragraph values in HTML for the <p></p> tags are such that the first line of a paragraph is not indented, so here we have made an adjustment using CSS to make the paragraphs look more like conventional paragraphs; the first line is indented 25 pixels. The screenshot of the example Tim Berners-Lee Web site in Figure 6.10 shows this style of paragraph.

The "1.0em" font size means that the paragraph text will equal the default size in the browser used to view the Web site. If the browser's default font size is 11 points, the "1.0em" size will be 11 points. (The "em" value is described in the section on "CSS and Font Properties" in Chapter 3.) Many other CSS selectors in a style sheet may use some variant of 1.0em. For example, if we use the value "1.2em" value for the <h4> or "heading 4" selector in the CSS code, the <h4> font will be 1.2 times the default text size in the user's browser.

Unless a different value is chosen for other selectors with a "p" in them—p. leftpquote, p.breadcrumbs, p.display-field, p.block—the values of the CSS code for paragraph text will cascade down and govern these types of paragraph. For example, their text will be in Calibri, Optima, Arial, or sans-serif font.

As an exercise, change the font family in your style sheet to Times New Roman and review the entire Web site. You can also manipulate the other values. Any changes in these CSS values will affect the look of the entire Web site.

Inserting pull quotes

"Pull quotes" are used to replicate what is known in the print world as a "callout"—a small area on a page that "calls out" something already in the body of text that surrounds it. You should only use pull quotes for memorable or pithy quotes, or ideas that you really want to draw the reader's attention to. For pull quotes to be effective they should be used sparingly, and the words in them should be chosen carefully; oftentimes this is the first thing, perhaps sometimes the only thing, a reader will read on the particular Web page.

This is the pull quote we will design here as an example:

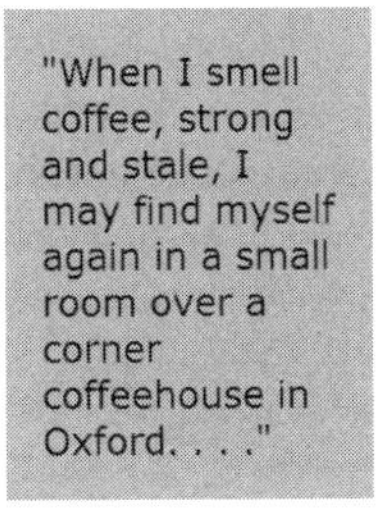

"When I smell coffee, strong and stale, I may find myself again in a small room over a corner coffeehouse in Oxford. . . ."

Figure 6.8 Pull quote.

A well-thought-out callout can draw the reader in; he or she may read the rest of your text to better understand what you have put in the callout.

In the CSS code below we are describing two "pquote" selectors for pull quotes—one for placing them to the right of the text and one for placing them to the left—but connecting them to the paragraphs, or "p"s, that precede them in the CSS file:

```
p.leftpquote {float: left; width: 8.0em; background: #D6C299;
padding: 1.0em; margin: 0.4em 0.4em 0.0em 0.0em;
font-size: 1.1em;}
p.rightpquote {float: right; width: 8.0em; background: #D6C299;
padding: 1.0em; margin: 0.4em 0.0em 0.0em 0.4em;
font-size: 1.1em;}
```

What this code does is produce a blocked-off area on the screen with a background color different than the white background of this particular Web site. It floats the pull quote to either the left or the right of the main text, and it allows for some padding around the text that makes up the pull quote. Additionally, it designates some margin spacing around the outside of the pull quote and describes the overall width. The margins for the left pull quote are 0.4em at the top and right, and for the right pull quote they are 0.4em at the top and left.

In the HTML file we use the following code where we want the pull quote to go:

<p class="leftpquote">"When I smell coffee, strong and stale, I may find myself again in a small room over a corner coffeehouse in Oxford. ..."</p>

We are using CSS classes here, and in this case we are embedding the code inside a <p> tag that has its values prescribed by the CSS code presented in the preceding subsection. This means that all of the stylistic elements which are *not* described in the pull quote CSS selectors p.leftpquote and p.rightpquote but which *are* described in the CSS for the paragraph styling will carry over, or "cascade" down, and be the operative style in the pull quotes.

Thus the font family and text indent values for p.leftpquote and p.right pquote will be the same as the CSS code for "p" in the previous subsection. However, the font size value for the p.leftpquote and p.rightpquote text will be 1.1em, slightly larger than the 1.0em value in regular paragraphs. The margins will also be different.

Producing hanging indents using CSS

Review Figure 6.13 later in this chapter, which shows the Works Cited page of the Tim Berners-Lee example Web site. Here is the HTML for one of the citations:

```
<p class="display-field">Berners-Lee, Tim. "Long Live the Web."
<i>Scientific American</i>. Dec. 2010: 80-85. Print.</p>
```

The class is "display-field" and it is part of the selector "p.display-field" in the following CSS code:

```
p.display-field { display: inline-block; padding-left: 50px; text-indent: -50px; }
```

What gives the hanging indent look to each citation is the "text-indent: -50px" declaration, which affects only the first line, the text moving to the left because of the negative value. Additionally, the "padding-left: 50px" declaration moves the text block to the right, overall. This distinguishes this text from the other paragraph types on the Web site.

Inserting an image with a caption using division tags

One way of inserting images with captions in a Web page is to use <div> tags. Because we are using the block-level element <div> tag to make a place for an image and its caption, we cannot embed it within a <p> tag, as this is also a block-level element; we also have to treat the tables in the same way as described in the next subsection. Below is the HTML code for this:

```
<div class="imageleftcaption" style="width:110px;">

<a href="http://www.w3.org">
<img src="w3c_homelarge.gif" width="110px" height="73px"
alt="W3C Logo"/>
</a>

W3C Logo

</div>
```

In this case, we are inserting an image of the W3C logo. If we use the W3C logo on a Web site, the W3C requires us to link it to its home page. This is why in the example shown the image source tag is surrounded by the <a href></a> tags that turn it into an image link.

Note that here we have designated that the width of this particular image is 110 pixels. If we had a different image with a different width we would have to adjust this. Also, we designate the CSS selector with class="imageleftcaption".

Using a CSS selector that we entitle ".imageleftcaption", we can determine the font size and family, the text alignment, where we will float the image, the margins around it, and its relative position:

> .imageleftcaption {font-size: 80%; text-align: center; float: left; margin: 0px 10px 0px 0px; position:relative; top:10px; font-family: Calibri, Optima, Arial, sans-serif;}

If we wanted the option of being able to float this image to the right, we would need this additional selector in our CSS file:

> .imagerightcaption {font-size: 80%; text-align: center; float: right; margin: 0px 0px 0px 10px; position: relative; top: 10px; font-family: Calibri, Optima, Arial, sans-serif;}

The only changes in this second CSS selector are that we have reduced the 10 pixel margin value that was designated for the right side of the image to 0 pixels, as we do not need this with the image on the right side, increased the left margin from 0 pixels to 10 pixels, and floated the image and caption to the right. Of course, we would also have to change our HTML to include class="imagerightcaption."

To see how the HTML code works, produce a page using <div class="image**right**caption" style="width:110px;"> and review it in a browser, then change the code to <div class="image**left**caption" style= "width:110px;"> and review how the image has been repositioned in the browser.

Next, in the .imageleftcaption CSS code, change the 10-pixel margin to 0 pixels. Then change the font size for the caption to another percentage value, or alter the text align value to "right" or "left." You might also change the relative positioning value to 5 or 15 pixels. After each change, review the file in a browser.

Inserting an image and caption using a table

If we are going to embed an image with a caption below it using a table design, we can use the following HTML code:

```
<table width="110px" class="imageleft">

<tr><td><a href="http://www.w3.org">
<img src="w3c_homelarge.gif" width="110" height="73"
alt="W3C logo"/></a></td></tr>

<tr><td class="caption">W3C Logo</td></tr>

</table>
```

Here we have one table with two rows, and in each row there is only one table division. The first row is for the image, and the second is for the caption.

Because we are using classes entitled "imageleft" or "imageright," and "caption" within our HTML <table> and <td> tags, we also need to add some lines to the CSS code:

```
table.imageleft {border: 0px; float: left; position: relative;
top: 5px;}

table.imageright {border: 0px; float: right; position: relative;
top: 5px;}

td.caption {font-family: Calibri, Optima, Arial, sans-serif;
font-size: 0.9em; text-align: center;}
```

The entire table has been defined using the CSS and HTML above. We have two choices of CSS selectors for our table, "imageleft" and "imageright." For example, if we want to position our table on the left we use the "table.imageleft" CSS code, and in the HTML we use <table width="110px" class="imageleft">. For the particular .gif image in this example, we need to indicate that the width should be 110 pixels. Other images to be placed on the Web site will have this value adjusted accordingly. For example, if we had another image that was, say, 205 pixels in width and we wanted to position it to the right of the text, we would use the tag <table width="205px" class="imageright">.

In the "table.imageleft" and "table.imageright" CSS selectors note that we have moved the tables down 5 pixels using relative positioning, given them borders of 0 pixels, and floated them to the left or right, as appropriate.

Additionally, we have used the CSS selector "td.caption" to state that the table cell, or "division," that the caption will appear in, <td></td>, will conform to the font family, font size, and text alignment specifications defined within the brackets in the CSS code. Note that we use the tag <td class="caption"> in the HTML code to apply the CSS code.

We have to insert the entire table *between* <p></p> tags, not *within* them. This is because in HTML 4.01 a block-level element cannot be embedded within another block-level element; a table cannot be embedded within a paragraph.

To see how the HTML code works, use it to produce a page, view it in a browser, then change the <table width="110px" class="imageleft"> to <table width="110px" class="imageright"> and see if it moves the image.

To test the CSS code, alter the text-align value for the caption from "center" to "right" or "left," or use another value than 0.9em for the font size, or change the relative positioning value for the tables, "position: relative," from 5 pixels to 10 or 15 pixels. You can also make other manipulations.

Because we have put much of our coding in CSS syntax, we can change many of the values on our entire Web site by manipulating just one file, the CSS file.

Sample informational web site

In this section we will detail a model informational Web site that takes as its subject Tim Berners-Lee, and which is represented in the hierarchical chart in Figure 6.7. Note that screenshots and code from this Web site also appear elsewhere in this book.

Review Figure 4.1, which is the Web site's home page, and then review the HTML/CSS code for it that follows (note that some of the text has been elided):

```
<!DOCTYPE html PUBLIC "-//W3C//DTD XHTML 1.0 Transitional//
EN"
"http://www.w3.org/TR/xhtml1/DTD/xhtml1-transitional.dtd">
<html xmlns="http://www.w3.org/1999/xhtml">

<head>
Tim Berners-Lee
<meta http-equiv="Content-Type" content="text/html;
charset=iso-8859-1" />

<!--The image of Tim Berners-Lee is used with the permission
of the W3C, and the W3C logo is used in accordance with this
```

organization's permission practices. All other information copyright the author. Both the HTML/CSS coding and the writing and research have been done by the author.-->

```
<meta name="description" content="The early career of Tim
Berners-Lee and the invention of the Internet, URIs, technolo-
gies of communication in place throughout history and how
each of them transfers and restructures information and
meaning, and the potential for invasion of privacy and abuse on
the WWW."/>
<meta name="keywords" content="Tim Berners-Lee, Berners-
Lee, Sir Tim Berners-Lee, URIs, URLs, inventor, invention, inventor
of the Internet, Internet, SGML, HTML, URIs, protocols, HTTP,
CERN, W3C, World Wide Web Consortium, privacy, abuse, open
source"/>

<meta name="author" content="J.D. Applen"/>

<link rel="stylesheet" type="text/css"
href="TBLbreadcrumbs.css"/>
</head>

<body>
<div id="container">

<div id="header"><img src="TBLimages/smalltimbernerslee.jpg"
height="131" width="200" align="left" alt="Tim Berners-
Lee"/><h1 style="position: relative; top: 30px">Tim Berners-
Lee</h1></div>

<div id="menu">
<ul>
<li><a href="home.html">Home</a></li>
<li><a href="inventor.html">Inventor</a></li>
<li><a href="futureoftheinternet.html">Future of the Internet
</a></li>
<li><a href="glossary.html">Glossary</a></li>
<li><a href="workcited.html">Works Cited</a></li>
</ul>
</div>
```

```
<div id="main-content">

<h4>Introduction to Tim Berners-Lee</h4>

<p>The life of Tim Berners-Lee is one that few people
know about. ... We all know who Bill Gates is and we also
know what Steve Jobs gave us; they produced the software
and personal computers that we use to connect to the
<a href="glossary.html#link9">Internet</a>. However, they
did not "invent" the Internet.</p>
<p class="block" style="position: relative; top: 20px;">
<a href="inventor.html">Inventor</a>    
Tim Berners-Lee invented HTML; URIs, which eventually became
URLS; browsers; the first server, and HTTP protocols; but he had
a problem with patents.</p>

<p class="block" style="position: relative; top: 20px;">
<a href="futureoftheinternet.html">Future of the Internet
</a>     The World Wide Web
Consortium (W3C), separation of layers, open source software,
and the potential for the abuse of individual privacy rights by
companies and governments.</p>
</div>

<div id="footer"><p><h6>Copyright &copy; 2011 by J.D. Applen.
All rights reserved.</h6></p></div>

</div>
</body>

</html>
```

Between the <head></head> tags are, among other things, the author's comments, three metatags, and the link to the external CSS file "TBLbreadcrumbs.css," shown here in bold.

Also note the major division IDs that determine the layout for the container, header, menu, main content, and footer. These divisions, and the division for the breadcrumbs that appear on the site, are described in the CSS code from the external CSS file, "TBLbreadcrumbs.css":

```
#container {
width: 1000px;
background-color: #ffffff;
margin: 0px auto;
}

#header {
text-align: center;
background-color: #D6C299;
height: 131px;
width: 1000px;
}

/*HTML5 does not support the inline <center> tags, so it is best
to get into the habit of using the "text-align: center;" code, as
above, in CSS*/

#menu {
float: left;
width: 1000px;
background-color: #e0e0e0;
font-family: Verdana, Arial, Helvetica, sans-serif;
color: 000000;
font-size: 1.0em;
}

#menu ul {
margin:0;
padding:0;
}

#menu ul li {
list-style: none;
display: inline;
}

/*"menu ul li" above removes the bullets.*/

#menu li a {
float: left;
padding: 10px 30px 10px 30px;
border-right: 1px solid #000000;
}
```

```
/*"menu ul a" above sets up the navigation block links. The "border-
right" property sets up a black line border to the right of each link.*/

#breadcrumbs {
padding: 10px 30px 10px 0px;
font-size: 0.9em;
font-family: Calibri, Optima, Arial, sans-serif;
position: relative;
right:-30px;
}

#main-content {
clear: left;
width: 900px;
height: auto;
padding: 15px 0px 0px 25px;
}

#footer {
height: 20pt;
font-size: 10pt;
width: 1000px;
background-color: #D6C299;
text-align: center;
margin:0px auto;
}

body {background-color: #b0c4de;}

/*This changes the background color outside the container. In
this case, it provides a nice contrast with the white background of
the main content area and area to the left and right of the main
content area.*/

p {font-family: Calibri, Optima, Arial, sans-serif; text-indent: 25px;
margin: 5px 30px 0px 0px; font-size: 1.0em;}

/*This code is for indenting paragraphs. */

p.block {margin-left: 50px; margin-right: 20px; text-indent: 0px;}

p.leftpquote {float: left; width: 8.0em; background: #D6C299;
padding: 1.0em; margin: 0.4em 0.4em 0.0em 0.0em; font-size:
1.1em;}
```

```
p.rightpquote {float: right; width: 8.0em; background: #D6C299;
padding: 1.0em; margin: 0.4em 0.0em 0.0em 0.4em;
font-size: 1.1em;}

/*"pquote" is for "pull quote" or callouts.*/

p.display-field {display: inline-block; padding-left: 50px; text-
indent: -50px;}

/*This code is used for the hanging indents in the Works Cited
page.*/

.imageleftcaption {font-size: 80%; text-align: center; float: left;
margin: 0px 10px 0px 0px; position: relative; top: 5px;
font-family: Calibri, Optima, Arial, sans-serif;}

.imagerightcaption {font-size: 80%; text-align: center; float: right;
margin: 0px 0px 0px 0px; position: relative; top: 5px;
font-family: Calibri, Optima, Arial, sans-serif;}

/*These two lines of code are for placing captions under images.*/

h1,h2,h3,h4,h5,h6 {font-family: Verdana, Arial, Helvetica, sans-serif;}

h4 {font-size: 1.2em;}
h5 {font-size: 1.1em;}
h6 {font-size: 0.8em; margin-bottom: 10px;}

a:link {text-decoration: none; color: #3300ff; font-size: 0.9em;
font-family: Verdana, Arial, Helvetica, sans-serif;}

a:visited {text-decoration: none; color: #3399ff; font-size: 0.9em;
font-family: Verdana, Arial, Helvetica, sans-serif;}

a:hover {text-decoration: underline; color: #ee99bb;
font-size: 0.9em; font-family: Verdana, Arial, Helvetica, sans-serif;}

a:active {text-decoration: underline; color: #000000;
font-size: 0.9em; font-family: Verdana, Arial, Helvetica, sans-serif;}

.italic {font-style: italic;}
```

The basics of major layout division IDs are described elsewhere, but it would be useful for you to go through each of the divisions in this code to study how they are used to set up the Web site. Contrast the properties and values in them with the code in the sections on CSS and layout in Chapter 3. As you can see, we have added a breadcrumbs ID that did not appear in those sections.

Many comments have been included in the /* ... */ code, and reviewing them will help you understand how this CSS code works. Many of the specifics are detailed in the previous section on "Additional HTML/CSS Code." Pay special attention to the comment that explains the background color of this Web site in the CSS coding for the "body." View Figures 4.1 and 4.4 to see how the background color serves as a contrast to the white background of the main content area.

Here is the HTML code for the file shown in Figure 4.4 (note that some code, for elements identical to those on the home page, and some text has been elided):

```
<!DOCTYPE html PUBLIC "-//W3C//DTD XHTML 1.0 Transitional//
EN"
"http://www.w3.org/TR/xhtml1/DTD/xhtml1-transitional.dtd">
<html xmlns="http://www.w3.org/1999/xhtml">

<head>
Tim Berners-Lee
<meta http-equiv="Content-Type" content="text/html;
charset=iso-8859-1" />
<link rel="stylesheet" type="text/css" href="TBLbreadcrumbs.
css"/>
</head>

<body>
<div id="container"> ...
<div id="header"> ... </div>
<div id="menu"> ... </div>
<div id="breadcrumbs"><a href="home.html">Home</a>
&rarr; <a href="inventor.html">Inventor</a> &rarr; <a
href="additionaltechnologies.html"><b>Additional
Technologies</b></a></div>

<div id="main-content">

<h5>Additional Technologies</h5>
```

```
<p>One of the virtues of the technologies that Berners-Lee devel-
oped was that all one had to do to be able to share information
with others was upload it to a server on the Web, and browsers
would translate it. To do this, and in addition to HTML, Berners-
Lee needed to invent the following technologies:</p>

<p class="block"><b>URIs</b>—&ldquo;Uniform resource
identifiers &rdquo; or URIs, which we now call uniform resource
locators or URLs, are the addresses for Web sites that allow us
to find a site on a server and review it without having to write
complex computer programs to find and access the information
on other computers (Berners-Lee and Fischetti 60). </p>

<p class="block"><b>Browser<b> ... <p>
<p class="block"><b>Web server<b> ... <p>
<p class="block"><b>HTTP<b> ... <p>

<p>With these URIs in place, the HTTP to help transfer the
information from the server to an individual's browser on
their computer, and the HTML used to format the text and other
information on the sites, everyone could post and share their
ideas and data. </p>

<p class="block" style="position: relative; top: 20px;">
<a href="patents.html">Patents and the WWW</a></p>

</div>

<div id="footer"><h6>Copyright &copy; 2011 J.D. Applen. All
rights reserved.</h6></div>

</div>
</body>
</html>
```

Note the four technologies listed are formatted by the <p class="block"> code. Note also the breadcrumbs division in bold, which is unique to this page and should be reviewed. Go over the section on "CSS Layout and Breadcrumbs" in Chapter 3 to make sure you understand this. Here is a screenshot of the breadcrumbs this code produces:

Figure 6.9 Screenshot of breadcrumbs in the Additional Technologies file.

The following screenshot shows a content page, headed "Future of the Internet," which features an image link with a caption, anchor links, and a breadcrumb structure specific to the page.

Figure 6.10 "Future of the Internet" page.

This is the HTML/CSS code for the breadcrumb and main content division IDs on this page, also the division class for "imageleftcaption" (note that one paragraph text has been elided, and that the other coding—for the header, menu, and so on—is the same as in previous files):

```
<div id="breadcrumbs"><a href="home.html">Home</a> &rarr;
<a href="futureoftheinternet.html"><b>Future of the Internet
</b></a></div>

<div id="main-content">

<h4>Future of the Internet</h4>

<p>The Internet would never have grown if everyone just
consumed information and did not produce it. For this to happen,
there could be no central control of the World Wide Web, so that
anyone who wanted to put information online could. Imagine if
there was only one server or one Internet browser, controlled by
```

a person or group of people who decided what constituted the Web and who had access to it. </p>

```
<div class="imageleftcaption" style="width: 110px;">
<a href="http://www.w3.org">
<img src="TBLimages/w3c_homelarge.gif" width="110px"
height="73px" alt="W3C Logo"/></a>
W3C Logo
</div>
```

<p>Beyond enabling people to contribute information to the <a href="glossary.html#link10">World Wide Web</a>, new technologies would have to be developed that allowed it to expand, and Berners-Lee hoped that more people would have a voice in deciding what new HTML and other related technological standards would be. To this end, he started the World Wide Web Consortium, also known as the W3C, and this is still the most important organization for unifying the many parties that have a stake in the development of the Web.</p>

<h6>HTML and Standardization</h6>

<p>While Berners-Lee invented HTML, HTML and other associated protocol languages are not owned by anyone. They are standards that the W3C consortium tries to get professionals from corporations and non-profit organizations to adopt for the technologies they will eventually design and manufacture, and they need to be updated from time to time. For example, browsers only work if they use HTML as this is the basis of the Web, the <a href="glossary.html#link8">lingua franca</a> of Web sites, and when there is a problem with some aspects of HTML, the people who manufacture browsers need to suggest new ideas, perhaps propose new HTML tags that replace older ones that are ineffective, and then agree to compromise on some standards, and then actually use these standards so the WWW works smoothly.</p>

<p> ... </p>

<p><a href="openstandards.html" style="position: relative; top: 20px;">Open Standards</a></p>

<p><a href="privacyandabuse.html" style="position:relative; top: 20px;">Privacy and Abuse</a></p>

</div>

On this page we have a graphic that is not only styled by the "imageleft-caption" CSS, it is an image link that, when selected, leads to the W3C Web site. Additionally, several anchor links to the Glossary page have been embedded in the text.

Here is the Glossary page for the Tim Berners-Lee Web site:

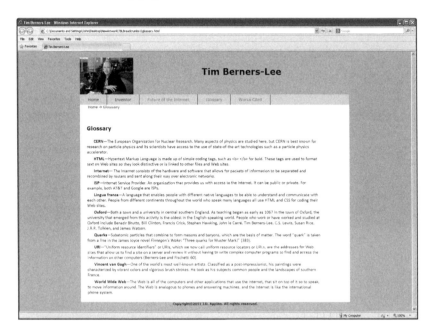

Figure 6.11 Glossary page.

The following code shows the first two of the ten glossary entries, and the breadcrumbs; the anchor links and breadcrumbs are shown in bold:

```
<div id="breadcrumbs"><a href="home.html">Home</a> &rarr;
<a href="glossary.html"><b>Glossary</b></a></p></div>

<div id="main-content">

<h4>Glossary</h4>

<p><b><a name="link1">CERN</a></b>—The European
Organization for Nuclear Research. Many aspects of physics are
studied here, but CERN is best known for research on particle
physics and its scientists have access to the use of state-of-the-
art technologies such as a particle physics accelerator.</p>
```

```
<p><b><a name="link2">HTML</a></b>—Hypertext Markup
Language is made up of simple coding tags, such as &lt;b&gt;
&lt;/b&gt; for bold. These tags are used to format text on Web
sites so it looks distinctive or is linked to other files and Web
sites.</p>
```

In the following screenshot of the "Inventor of HTML" page are two "pull quotes" or "callouts," one of which was detailed in the "Additional HTML/CSS Code" section earlier in this chapter:

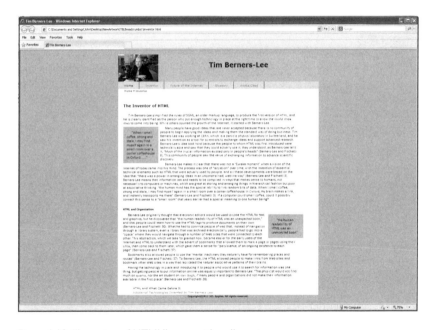

Figure 6.12 "Inventor of HTML" page with pull quotes.

Note how the paragraph text flows around the callouts, and how they are positioned relative to the left and right sides of the page. Below is the code for this page (note that, again, some text has been elided, and that the other coding is the same as in previous files):

```
<div id="breadcrumbs"><a href="home.html">Home</a> &rarr;
<a href="inventor.html"><b>Inventor</b></a></div>

<div id="main-content">
```

\<h4\>The Inventor of HTML\</h4\>

\<p\>Tim Berners-Lee simplified the rules of SGML, an older markup language, to produce the first version of \<a href="glossary. html#link2"\>HTML\</a\>, and he is clearly identified as the person who put enough technology in place at the right time to allow the \<a href="glossary.html#link10"\>World Wide Web\</a\> to come into being. While others spurred the growth of the Internet, it started with Berners-Lee.\</p\>

\<p class="leftpquote"\>"When I smell coffee, strong and stale, I may find myself again in a small room over a corner coffeehouse in Oxford. ..."\</p\>

\<p\>Many people have good ideas that are never accepted because there is no community of people to begin applying the ideas and making them the standard way of doing business. Berners-Lee was working at \<a href="glossary.html#link1"\>CERN \</a\>, which is a particle physics laboratory in Switzerland, and he saw his invention as a tool for scientists to exchange ideas and support advanced research. Berners-Lee's idea took hold because the people to whom HTML was first introduced were technically able and they saw that they could actually use it; they understood, as Berners-Lee tells it, "Much of the crucial information existed only in people's heads" (Berners-Lee and Fischetti 9). This community of people saw the value of exchanging information to advance scientific discovery.\</p\>

\<p\> ... \</p\>

\<h6\>HTML and Organization\</h6\>

\<p class="rightpquote"\>"the human readability of HTML was an unexpected boon"\</p\>

\<p\>Berners-Lee originally thought that electronic editors would be used to code the HTML for text and graphics, but he discovered that "the human readability of HTML was an unexpected boon," and that people could learn how to use the HTML tags to produce documents on their own (Berners-Lee and Fischetti 90). However, what he had to convince people of was that, instead of navigation through a library system, even a library that was archived electronically, people had to go into a "space" where they would navigate through a number of Web sites that were connected to each other. This abstraction, which we take for granted now, became easier for the early users of the Internet and HTML to understand with the advent of bookmarks that allowed them to mark a page or pages using their <a href="glossary.html#link4">URIs</a>, then come back to them later, which gave them a sense of "persistence, of an ongoing existence to each page" (Berners-Lee and Fischetti 37). </p>

<p> ... </p>

<p class="block" style="position: relative; top: 20px;">
<a href="html.html">HTML and What Came Before It</a></p>

<p class="block" style="position: relative; top: 20px;">
<a href="additionaltechnologies.html">Additional Technologies Invented by Tim Berners-Lee</a></p>

</div>

It is important to see how the callouts are positioned relative to the paragraphs here. Because they are block-level elements, they must be between the paragraphs, not within them.

The Works Cited page for the Tim Berners-Lee Web site is shown here:

Figure 6.13 "Works Cited" page.

The following code provides for the unique formatting of the Work Cited file:

```
<div id="breadcrumbs"><a href="home.html">Home</a> &rarr;
<a href="workcited.html"><b>Works Cited</b></a></div>

<div id="main-content">

<p><h5><div style="text-align: center">Works Cited</div>
</h5></p>

<p class="display-field">Berners-Lee, Tim. "Long Live the Web."
<i>Scientific American</i>. Dec. 2010: 80-85. Print.</p>

<p class="display-field">Berners-Lee, Tim, and Mark Fischetti.
<i>Weaving the Web: The Original Design and Ultimate Destiny
of the World Wide Web by its Inventor.</i> San Francisco, CA:
Harper Collins, 1999. Print.</p>

<p class="display-field">Gleick, James. "Patently Absurd."
<i>New York Times Magazine</i>. 11 Oct. 2010.
&lsaquo;http:www.nytimes.com/2000/03/12/magazine/
patently-absurd.html?scp=1&sq=jeff bezos patent barnes and
noble amazon&st=cse.&rsaquo; Web.</p>

<p class="display-field">Joyce, James. <i>Finnegan's Wake</i>.
New York: Penguin Classics, 1999. Print.</p>
```

Note how the code "text-align: center" affects the page head, and the effect of the "display-field" class that indents the second lines of the references listed.

Drop-down menus

In this section, we will alter the sample Web site to change it from using a breadcrumbs navigation format to using a drop-down menu format that you may prefer to use for your own informational Web site.

Review the difference between the breadcrumbs on the "Future of the Internet" page shown in Figure 6.10 and the drop-down menu shown here:

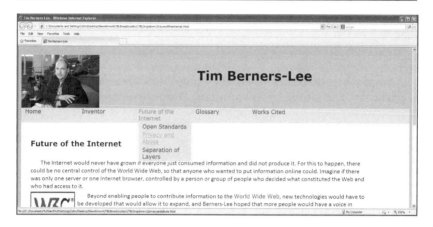

Figure 6.14 Drop-down menu.

Additionally, to better show off the strengths of drop-down menus, we are changing the navigation structure so that we have two sets of second-level drop-down links, each with three links, as reflected in this adjusted hierarchy:

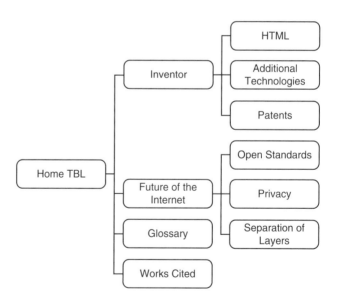

Figure 6.15 Tim Berners-Lee Web site hierarchy adjusted for drop-down menus.

The first-level links are "Home," "Inventor," "Future of the Internet," "Glossary," and "Works Cited." Note that we are using the term "first-level" here, not "first-tier," as it describes something different—the links across the top of the page in the menu bar, to the home page, "Home TBL," and the "Inventor," "Future of the Internet," "Glossary," and "Works Cited" files. The "first-tier" links do not include "Home TBL."

The second-level links reside "below" the "Inventor" and "Future of the Internet" first-level links.

To understand what needs to be done to make this change, we will review the differences in the CSS, then in the HTML.

The following code is from the external CSS file "TBLbreadcrumbs.css" discussed previously, and is for the "#menu" ID and "a:hover" pseudo-class code:

```
#menu {
float: left;
width: 1000px;
background-color: #e0e0e0;
font-family: Verdana, Arial, Helvetica, sans-serif;
color: 000000;
font-size: 1.0em;
}

a:hover {text-decoration: underline; color: #ee99bb;
font-size: 0.9em; font-family: Verdana, Arial, Helvetica, sans-serif;}
```

For the drop-down menus to work, we need to replace this code with the following:

```
#menu {
float: left;
width: 1000px;
background-color: #e0e0e0;
font-family: Verdana, Arial, Helvetica, sans-serif;
color: 000000;
font-size: 1.0em;
height: auto;
}

#menu ul {
display: block;
```

```
margin: 0px;
padding: 0px;
list-style: none;
}

#menu a {
display: block;
width: 8.5em;
padding: 0px 0px 4px 0px;
}

#menu li {
float: left;
width: 8.5em;
padding: 0px 0px 0px 10px;
}

#menu li ul {
display: none;
}

#menu li:hover ul {
display: block;
width: 8.5em;
position: absolute;
background-color: #e0e0e0;
}

a:hover {text-decoration: underline; color: #ee99bb;
font-size: 0.9em; font-family: Verdana, Arial, Helvetica, sans-serif;
background-color: #e0e0e0;}
```

Here are the explanations of how these menu IDs work:

- **#menu**—In the basic menu ID, we have added "height: auto;" to the CSS (shown in bold). This enables us to see the entire vertical expanse of the secondary links that drop down. This code needs to be in place for some browsers.
- **#menu ul**—The "display: block" declaration makes the entire rectangular area surrounding the text a link; you can place the cursor anywhere in this area, not just over the text, and select the link.

The "list-style: none" keeps the default bullets from displaying. There are 0 pixels of padding around the text.

- **#menu a**—The "8.5em" describes the width of the link block in the menu. The padding values mean there are 4 pixels of space below each text link in the menu bar. If you increase or decrease this value you will see that the links move farther apart or closer together.

- **#menu li**—The "float: left" positions the first-level links horizontally across the menu area on the screen. If we did not put this in, the links would stack up vertically, one above the other. The padding gives 10 pixels of space on the left of the text for each link in the menu bar. If you increase or decrease this value, you will see that the links shift right or left.

- **#menu li ul**—The "display: none" declaration relates to the second-level list that, in the example shown in Figure 6.14, starts with "Open Standards." If this was not included, both the first- and second-level links would show up. The second-level links only show up when, in this case, the "Future of the Internet" link is hovered over.

- **#menu li:hover ul**—The "position: absolute" declaration here keeps the drop-down background color that shows up in the second-level links field from spreading left and right. We are also giving the background to the second-level links the same color as that of the first-level links, in this case "#e0e0e0."

We also alter the CSS for the "a:hover" pseudo-class code by adding "background-color: #e0e0e0;" to it (shown in bold). This means that the background of the second-level links will be visible when a first-level link is hovered over, in all browsers.

Now that we have changed the style with the CSS code, we need to set up the actual structure of the links using new HTML code (note that some of the code which is the same as previously has been elided):

```
<link rel="stylesheet" type="text/css" href="dropdownTBL.css"/>
</head>

<body>
<div id="container">

<div id="header"><img src="TBLimages/smalltimbernerslee.jpg"
height="131" width="200" align="left" alt="Tim Berners-
Lee"/><h1 style="position:relative; top:30px">Tim Berners-
Lee</h1></div>
```

```
<div id="menu">
<ul>
<li><a href="home.html">Home</a></li>

<li><a href="inventor.html">Inventor</a>
    <ul>
    <li><a href="help.html">HTML</a></li>
    <li><a href="help.html">Additional Technologies</a></li>
    <li><a href="patents.html">Patents</a></li>
    </ul>
</li>

<li><a href="futureoftheinternet.html">Future of the Internet
</a>
    <ul>
    <li><a href="openstandards.html">Open Standards</a></li>
    <li><a href="privacyandabuse.html">Privacy and Abuse
    </a></li>
    <li><a href="separationoflayers.html">Separation of Layers
    </a></li>
    </ul>
</li>

<li><a href="glossary.html">Glossary</a></li>

<li><a href="workcited.html">Works Cited</a></li>
</ul>
</div>

<div id="main-content">

<h4>Future of the Internet</h4>

<p>The Internet would never have grown if everyone just con-
sumed information and did not produce it ...

[...]

<p><a href="privacyandabuse.html" style="position:relative;
top: 20px;">Privacy and Abuse</a></p>

</div>
```

```
<div id="footer"><h6>&copy; 2011 J.D. Applen. All rights
reserved.</h6></div>

</div>
</body>
```

The HTML for the "breadcrumb" divisions, or <div></div> tags, has been removed and a new menu division added, shown here in bold. We have also changed to a new external CSS file, "dropdownTBL.css," again shown here in bold, which includes the changes to the CSS described.

Works cited

Mayo Clinic. Mayo Foundation for Medical Education and Research, 2012. Web. March 15, 2012.

Appendix

In the following letter, Sullivan Ballou writes to his wife Sarah. He was a Union soldier during the American Civil War, and the letter was written one week before his death in the Battle of Bull Run, July 21, 1861. It has been abridged and is reprinted with the permission of the Rhode Island Historical Society.

Head-Quarters, Camp Clark
Washington, D.C., July 14, 1861

My Very Dear Wife:

Indications are very strong that we shall move in a few days, perhaps tomorrow. Lest I should not be able to write you again, I feel impelled to write a few lines, that may fall under your eye when I shall be no more.

I have no misgivings about, or lack of confidence in, the cause in which I am engaged, and my courage does not halt or falter. I know how strongly American civilization now leans upon the triumph of government, and how great a debt we owe to those who went before us through the blood and suffering of the Revolution, and I am willing, perfectly willing to lay down all my joys in this life to help maintain this government, and to pay that debt. ...

Sarah, my love for you is deathless. It seems to bind me with mighty cables, that nothing but Omnipotence can break; and yet, my love of country comes over me like a strong wind, and bears me irresistably on with all those chains, to the battlefield. The memories of all the blissful moments I have spent with you come crowding over me, and I feel most deeply grateful to God and you, that I have enjoyed them so long. And how hard it is for me to give them up, and burn to ashes the hopes of future years, when, God willing, we might still have lived and loved together, and seen our boys grow up to honorable manhood around us.

I know I have but few claims upon Divine Providence, but something whispers to me, perhaps it is the wafted prayer of my little Edgar, that I shall return to my loved ones unharmed. If I do not, my dear Sarah, never forget how much I love you, nor that, when my last breath escapes me on the battlefield, it will whisper your name.

Forgive my many faults, and the many pains I have caused you. How thoughtless, how foolish I have oftentimes been! How gladly would I wash out with my tears, every little spot upon your happiness, and struggle with all the misfortune of this world, to shield you and my children from harm.
. . .

But, O Sarah, if the dead can come back to this earth, and flit unseen around those they loved, I shall always be near you—in the garish day, and the darkest night amidst your happiest scenes and gloomiest hours—always, always, and, if the soft breeze fans your cheek, it shall be my breath; or the cool air cools your throbbing temples, it shall be my spirit passing by. . . .

"Sullivan."

Index

Note: 'N' after a page number indicates a note; 'f' indicates a figure; 't' indicates a table.

grammar 215–21
graphics 58–63. *See also* image maps;
   images
Grusin, Richard 16

Havelock, Eric 7
Hawhee, Debra 181, 191
headers: CSS 92; definition of 97;
   function of 238–9. *See also* footers
headings: margins and justification for
   233; tags for creating 45–6
head tags, as component of XHTML 36
hierarchical structures 258–9
home pages: avoidance of jargon on 243;
   branding of 239–40, 242; linking to
   forms from 243; personality of 241;
   search engine boxes on 243; taglines
   on 240–2
Horton, Sarah 229, 231, 233, 238
HTML: adding comments to code 170;
   as component of XHTML 36;
   deprecated tags 48–9; HTML5 48–9;
   as human readable 30; and images
   136–8; indenting and spacing in
   234–5; origins of 28–34; as standard
   31; used to format online text 34–5;
   validating code 47–8. *See also* CSS;
   tags; XHTML
HTTP, definition of 30
Human Rights Watch 241
hypermediacy: definition of 16; of
   television programs 15; of Web sites 16
hypertext markup language. *See* HTML
hypertext structures 259
Hypertext Transmission Protocol.
   See HTTP
Hyundai Motor Company 240–1

iceberg syndrome 249
idioms 2–3
image maps: client- and server-side 58;
   definition of 58; tags for 58–9
images: alt tags for 136; backgrounds
   167–70; borders for 138–40; and
   captions 284–7; file formats 135;
   and HTML 136–8; links 140–2;
   padding around 151–4;
   resizing files 143; thumbnail
   links 143–5
indenting and spacing 234–5, 284
independent clauses 216
individuality 8–9, 21–2
information literacy xii

interactivity, myth of, for new media
   11–12
Internet: Berners-Lee's contributions to
   development of 28–34; definition of
   32; and privacy loss 33; standardization
   and infrastructure of 31–2; transferring
   files to, using
   FTP 174–6
Internet Service Providers. *See* ISPs
inverted pyramid technique 208–14
irony 184
ISPs 33
italics: CSS 92–3; tags for creating 38–9;
   when to use 230–1
iTunes 32–3

Jefferson, Thomas 192–4, 196
Jeopardy (television program) 23–4
Johnson, Robert 192
JPEG files 135. *See also* images

kairos: Obama's speech as example of
   186–8; as rhetorical situation 186;
   spatial element of 186–8; temporal
   element of 186–8
Kaplan, Robert 208–9
Krug, Steve 240, 248, 257, 260

language: computer 2; concrete vs.
   abstract 204–5; intertextuality of
   192–3; oral 2; slang 2–3; as socially
   constructed 3, 195–6; of users vs.
   developers 243. *See also* writing style
Lanham, Richard 7
late age of print 12–13
layouts: creating consistent patterns for
   231–2; indenting and spacing of
   234–5; margins and justification for
   233; screen length 235–8; white space
   232–5
letter writing 14–15, 308–9
line breaks, creating 38–9
line heights, changing, using CSS 88–91
links: absolute 53–5; alphabetic list of
   124–9, 246–7, 250–1; alternatives
   colors and styles for 160–2; anchor
   63–7; deep 55; for email addresses 58;
   on home pages 243; image maps
   58–63; images as 140–2; relative
   53–5; structuring 247–8; thumbnail
   143–5
lists, ordered, tags for creating 44–5
lists, unordered, tags for creating 42–4